PRIVACY AND IDENTITY PROTECTION

DRONES IN THE U.S.

PRIVACY ISSUES AND REGULATORY CONSIDERATIONS

PRIVACY AND IDENTITY PROTECTION

Additional books in this series can be found on Nova's website under the Series tab.

Additional e-books in this series can be found on Nova's website under the eBooks tab.

PRIVACY AND IDENTITY PROTECTION

DRONES IN THE U.S.

PRIVACY ISSUES AND REGULATORY CONSIDERATIONS

SOPHIA BARRETT
EDITOR

New York

NOTICE TO THE READER

Library of Congress Cataloging-in-Publication Data

ISBN: 978-1-63484-918-0

Published by Nova Science Publishers, Inc. † New York

CONTENTS

PREFACE

It has been three years since Congress enacted the FAA Modernization and Reform Act of 2012 (FMRA), calling for the integration of unmanned aircraft systems (UAS), or "drones," into the national airspace by September 2015. During that time, the substantive legal privacy framework relating to UAS on the federal level has remained relatively static: Congress has enacted no law explicitly regulating the potential privacy impacts of drone flights, the courts have had no occasion to rule on the constitutionality of drone surveillance, and the Federal Aviation Administration (FAA) did not include privacy provisions in its proposed rule on small UAS. This issue, however, has not left the national radar. Congress has held hearings and introduced legislation concerning the potential privacy implications of domestic drone use; President Obama recently issued a directive to all federal agencies to assess the privacy impact of their drone operations; and almost half the states have enacted some form of drone legislation. This book provides a primer on privacy issues related to various UAS operations, both public and private, including an overview of current UAS uses, the privacy interests implicated by these operations, and various potential approaches to UAS privacy regulation.

In: Drones in the U.S.
Editor: Sophia Barrett

ISBN: 978-1-63484-918-0

Chapter 1

DOMESTIC DRONES AND PRIVACY: A PRIMER

Richard M. Thompson II

SUMMARY

It has been three years since Congress enacted the FAA Modernization and Reform Act of 2012 (FMRA), calling for the integration of unmanned aircraft systems (UAS), or "drones," into the national airspace by September 2015. During that time, the substantive legal privacy framework relating to UAS on the federal level has remained relatively static: Congress has enacted no law explicitly regulating the potential privacy impacts of drone flights, the courts have had no occasion to rule on the constitutionality of drone surveillance, and the Federal Aviation Administration (FAA) did not include privacy provisions in its proposed rule on small UAS. This issue, however, has not left the national radar. Congress has held hearings and introduced legislation concerning the potential privacy implications of domestic drone use; President Obama recently issued a directive to all federal agencies to assess the privacy impact of their drone operations; and almost half the states have enacted some form of drone legislation.

There are two overarching privacy issues implicated by domestic drone use. The first is defining what "privacy" means in the context of aerial surveillance. Privacy is an ambiguous term that can mean different things in different contexts. This becomes readily apparent when attempting to apply traditional privacy concepts such as personal control and secrecy to drone surveillance. Other, more nuanced privacy theories such as personal autonomy and anonymity must be explored to get a fuller understanding of the privacy risks posed by drone surveillance. Moreover, with ever-increasing advances in data storage and manipulation, the subsequent aggregation, use, and retention of drone-obtained data may warrant an additional privacy impact analysis.

The second predominant issue is which entity should be responsible for regulating drones and privacy. As the final arbiter of the Constitution, the courts are naturally looked upon to provide at least the floor of privacy protection from UAS surveillance, but as will be discussed in this report, under current law, this protection may be minimal. In addition to the courts, the executive branch likely has a role to play in regulating privacy and drones. While the FAA has taken on a relatively passive role in such regulation, the President's new privacy directive for government drone use and multi-stakeholder process for private use could create an initial framework for privacy regulations. With its power over interstate commerce, Congress has the broadest authority to set national standards for UAS privacy regulation. Several measures were introduced in the 113th Congress that would have restricted both public- and private-actor domestic UAS operations, and reintroduction of these bills is likely in the 114th Congress. Lastly, some have argued that under our system of federalism, the states should be left to experiment with various privacy schemes. It is reported that by the end of 2014, 20 states have enacted some form of drone regulation.

This report will provide a primer on privacy issues related to various UAS operations, both public and private, including an overview of current UAS uses, the privacy interests implicated by these operations, and various potential approaches to UAS privacy regulation.

INTRODUCTION

It has been three years since Congress enacted the FAA Modernization and Reform Act of 2012 (FMRA), calling for the integration of unmanned aircraft systems (UAS), or "drones," into the national airspace by September 2015.[1] During that time, the substantive legal privacy framework relating to UAS on the federal level has remained relatively static: Congress has enacted no law explicitly regulating the potential privacy impacts of drone flights; the courts have had no occasion to rule on the constitutionality of drone surveillance; and the Federal Aviation Administration (FAA) did not include privacy provisions in its proposed rule on small UAS.[2] However, this issue has not left the national radar. Congress held several hearings in the 113th Congress on the potential privacy impacts of domestic drone use on American citizens;[3] Members have introduced legislation to regulate both public- and private-actor use of drones;[4] the executive branch has taken various measures to assess the privacy impact of its drone operations;[5] and almost half the states have enacted UAS legislation.[6]

As this report will discuss, there are two overarching privacy issues implicated by domestic drone flights. The first issue is defining and understanding what the chameleon phrase "privacy" means in the context of aerial surveillance. Traditional privacy concepts such as the right to control information about oneself or secrecy do not adequately capture potential privacy concerns raised by visual surveillance; instead, privacy concepts such as personal autonomy and anonymity must be explored to get a fuller understanding about the scope of privacy interests implicated by UAS operations. Additionally, a separate set of privacy interests might be implicated by the subsequent aggregation, use, and retention of drone-obtained information. For instance, the Supreme Court's aerial surveillance cases generally hold that it is not a Fourth Amendment search to conduct surveillance of private property while flying in navigable airspace.[7] However, one could argue that beyond the initial collection of data, a unique privacy interest is at risk by aggregating multiple flights over one's home, using drone-obtained data in ways never envisioned by the initial collection, or retaining that data indefinitely.

The second predominant issue is which entity should be responsible for regulating UAS privacy issues. The courts can be expected to apply traditional privacy rules encompassed in the Fourth Amendment's prohibition against unreasonable searches and privacy torts found largely in state statutory and common law. Since drone use has been relatively limited to date, the courts have yet to address how these laws apply to UAS flights. Privacy safeguards might also come from executive branch policies. On February 15, 2015, President Obama issued a memorandum charging all federal agencies that use drones in their operations to evaluate the privacy impact of such use and develop policies to mitigate any privacy concerns.[8] The need for this new initiative may have been prompted, in part, by the FAA's relatively passive role in setting privacy rules for drone operations. The FAA's recently proposed rule for small UAS operations and certifications did not include any privacy provisions, and it is likely the FAA will remain in an advisory, rather than regulatory, role with respect to this issue.[9] With its power over interstate commerce,[10] Congress has the broadest authority to set national standards for UAS privacy regulation. For instance, Congress could create a broad legislative scheme regulating all UAS uses; it could pass more narrow proposals such as a warrant requirement for law enforcement use; or it could create a scheme regulating the subsequent use, retention, and dissemination of drone-obtained data. Federal legislation introduced in the 113th Congress, and likely to be introduced in the 114th Congress, utilized, to some extent, these various approaches.[11] Lastly, some have argued that under our system of federalism, the states should be left to experiment with various privacy schemes.[12] Indeed, the states have taken up this call with 20 states reportedly enacting laws addressing UAS issues by the end of 2014.[13]

In reviewing these various forms of regulation, it is clear that understanding privacy rights vis-à-vis drones is not as simple as applying Supreme Court case law or federal and state statutes. Rather, regulations may come from myriad sources, some statutory, some regulatory, and some practical. With this in mind, this report will provide a primer on privacy issues relating to various UAS operations, both public and private, including an overview of current UAS uses, the privacy interests implicated by these operations, and various potential approaches to UAS privacy regulation.[14]

BACKGROUND

The FMRA defined an "unmanned aircraft" to mean "an aircraft that is operated without the possibility of direct human intervention from within or on the aircraft."[15] An "unmanned aircraft system" (UAS) is the unmanned aircraft and its "associated elements (including communications links and the components that control the unmanned aircraft) that are required for the pilot in command to operate safely and efficiently in the national airspace system."[16] UAS, commonly referred to as "drones," can range from the size of an insect—sometimes called nano or micro drones—to the size of a traditional jet.[17] Drones can be outfitted with an array of sensors, including high-powered cameras,[18] thermal imaging devices,[19] license plate readers,[20] and laser radar (LADAR).[21] In the near future, drones might be outfitted with facial recognition or soft biometric recognition, which can recognize and track individuals based on attributes such as height, age, gender, and skin color.[22] In addition to their sophisticated sensors, the technical capability of drones is rapidly advancing. For instance, the Defense Advanced Research Projects Agency (DARPA), the technology research arm of the U.S. military, is working on a drone that can enter a building through a window and fly at speeds up to 20 meters/second without communication to the operator and without GPS waypoints.[23] As discussed below, these advanced sensors and capabilities have the capacity to heighten privacy risks posed by drones.

Authorization and Uses

Currently, all UAS operators who do not fall within the recreational use exemption (discussed below) must apply directly to the FAA for permission to fly.[24] The process for obtaining permission to operate differs depending on whether the operator is a public operator or a private commercial operator.

Public Operators

UAS operated by federal, state, or local agencies must obtain a certificate of authorization or waiver (COA) from the FAA.[25] After receiving COA applications, which can be completed online, the FAA conducts a comprehensive operational and technical review of the UAS and can place limits on its operation in order to ensure its safe use in airspace.[26] COAs are not generally publicly available, but have been released in response to Freedom of Information Act (FOIA) requests. The most recent FOIA request on the FAA's website revealed 426 UAS COA files.[27]

Customs and Border Protection

The most prominent domestic user of UAS among federal agencies is the Department of Homeland Security's (DHS's) U.S. Customs and Border Protection (CBP).[28] As of September 2013, CBP was reported to own 10 UAS, which are operated by CBP's Office of Air and Marine (OAM).[29] The bulk of CBP's flight missions involve patrolling the nation's borders, interdicting persons and contraband illegally entering the United States. CBP's COA allows it to operate in an airspace that covers a 100 mile corridor along the northern border and within 20 to 60 miles along the southern border, excluding urban areas. In addition to its border missions, CBP has provided unmanned aerial support to various federal and state agencies, including the Drug Enforcement Administration (DEA), Immigration and Customs Enforcement (ICE), the U.S. Marshals Service, the U.S. Coast Guard, the Bureau of Land Management, and the Texas Department of Public Safety, among others.[30] In one prominent case, CBP used one of its Predator drones to assist local police in North Dakota to monitor an individual suspected of cattle theft and threatening police officers.[31] The number of these "loan" operations has steadily increased each year since 2010. For instance, CBP flew one flight for the DEA in 2010, 19 flights in 2011, and 66 in 2012.[32]

Department of Justice

The other primary federal agency currently using UAS in its operations is the Department of Justice (DOJ), the nation's chief law enforcement agency. In 2013, DOJ's Office of the Inspector General issued a report reviewing the various UAS programs underway within DOJ.[33] All of the UAS purchased by DOJ so far have been what the FAA calls "small UAS," those 55 pounds or less. The Federal Bureau of Investigation (FBI) has been the most prominent component of DOJ to use UAS in the field and has been doing so since 2006.[34] The FBI noted in a July 2013 letter to Senator Rand Paul that it had used UAS in 10 operations, including those related to search and rescue, drug interdictions, kidnapping, and fugitive investigations.[35] The Bureau of Alcohol, Tobacco, Firearms and Explosives (ATF) has plans to use UAS in its future operations but has not done so yet.[36] And while the DEA and the Marshals Service have purchased several UAS, as of 2012 they had no plans to use them in future operations.[37] This might be attributed to the DEA and Marshals Service use of CBP drones in their operations.

Other Federal Uses

In addition to DHS and DOJ, the FAA has issued COAs to various other federal entities—both military and civilian.[38] On the defense side, COAs have been issued to DARPA, the U.S. Army, the Navy, the Marine Corps, and the Air Force, for operations in U.S. airspace. On the civilian side, COAs have been issued to the Departments of State, Interior, and Energy, the National Aeronautics and Space Administration (NASA), and the National Institute of Standards and Technology (NIST).

State and Local Governmental Operations

Like their federal counterparts, state and local governmental entities must obtain a COA in order to conduct drone operations. As of 2012, the FAA has issued several hundred COAs to state and local governmental entities to operate UAS.[39] These have included state and city governments, such as Houston, Texas, and the Colorado Department of Transportation; local fire and police departments, including the Miami-Dade Police Department; and state and local universities, like Indiana University.[40]

Private Operators

Until the FAA finalizes its small UAS rule as required under FMRA, there are currently only a limited number of ways to gain authorization for private commercial use of UAS. A private operator may obtain a special airworthiness certificate in the experimental category issued by the FAA.[41] These certificates have been issued on a limited basis for flight tests, demonstrations, and training. The second means of authorization is under FMRA's Section 333, which permits the FAA to authorize certain UAS flights before the FAA issues a final small UAS rule.[42] These Section 333 exemptions have been issued for such purposes as movie productions, precision agriculture, flare stack inspection, and bridge inspections.[43] Finally, certain recreational UAS users may fall within FMRA's model aircraft exception, meaning they are not required to obtain specific authorization from the FAA before flying UAS. Section 336 of FMRA prohibits the FAA from promulgating a rule regarding "model" aircraft if the aircraft is flown "strictly for hobby or recreational use" and meets several other requirements.[44]

PRIVACY INTERESTS IMPLICATED BY UAS OPERATIONS

A threshold issue in analyzing drones and privacy is determining what privacy interests might be implicated by drone operations, both public and private. Privacy is an ambiguous term that can mean different things in different contexts, which becomes apparent when attempting to apply traditional privacy concepts to drone surveillance.

Surveillance

The first privacy interest implicated by the use of UAS is the initial collection of information about people—what can be called "surveillance"—whether it is conducted by a government or private actor. With respect to UAS operations, surveillance takes place in nearly all flights, as one of their major purposes is to collect information, namely visual surveillance, from the sky above. Surveillance might entail a broad and indiscriminate recording of people on the ground using a camera sensor on the aircraft. For instance, the U.S. Air Force recently rolled out the newest iteration of its "Gorgon Stare" sensor, a remotely controlled, aircraft-based Wide-Area Persistent Surveillance (WAPS) system.[45] This sensor allows a single UAS to monitor a

100km² area in high resolution for several hours at a time.[46] CBP uses a similar sensor while monitoring the U.S. border.[47] Alternatively, surveillance might consist of more targeted data collection, such as the FBI's use of a drone in 2013 to monitor a standoff with a child kidnapper.[48] Both types of surveillance, to one degree or another, implicate various privacy concepts, including personal control, secrecy, autonomy, and anonymity.

Personal Control

One of the leading privacy theories is the right to control information about oneself. According to one prominent privacy theorist, "[p]rivacy is the claim of individuals, groups, or institutions to determine for themselves when, how, and to what extent information about them is communicated to others."[49] This paradigm works in some mediums, for example, when we consent to a smart phone app tracking our location: we can decide whether the trade-off to our privacy is worth the service we will receive in exchange. However, with drones and aerial surveillance, however, this theory of privacy breaks down. For instance, should we be able to control whether people can view us or our activities in public? How are we expected to exercise such control? And should it make a difference if a person sees us from a traditional aircraft versus a drone? Requiring consent before conducting aerial surveillance could undermine many uses of this new technology, making this theory of privacy as applied to drone surveillance unworkable.

Secrecy

Another prominent theory of privacy is the secrecy model. Under this model, an individual's privacy is invaded if previously concealed information about them is publicly disclosed; or, put another way, an individual is not entitled to privacy where that information has been revealed to another person.[50] The Supreme Court relied on this secrecy model in the three aerial surveillance cases from the 1980s, which held that surveillance conducted in public airspace does not even trigger—let alone violate—the Fourth Amendment's protection against unreasonable searches.[51] Citing to an earlier case, the Court observed in *California v. Ciraolo* that "[w]hat a person knowingly exposes to the public, even in his own home or office, is not a subject of Fourth Amendment protection."[52] In addition to Fourth Amendment case law, the privacy tort intrusion upon seclusion applies this same secrecy model such that most activities revealed to the public would not constitute a violation of this tort.[53] If secrecy remains the primary model for the Fourth Amendment and privacy torts, individuals would have little protection from drone surveillance when their location and activities have been revealed to the public.

Autonomy

Another important theory of privacy is personal autonomy. Autonomy generally refers to the ability of an individual to make life decisions free from interference or control by both government and private actors.[54] Some have argued that surveillance represents a form of social control, mandating conformity in society, hindering independent thinking, and recording and notating people who stray from acceptable norms.[55] One such example is Jeremy Bentham's Panopiticon, a prison facility designed to give inmates the perception of being watched at all times causing them to self-regulate their behavior to the desired norm.[56] Of course, some form of social control—say, crime control—is beneficial and expected in every society. As noted by Professor Daniel Solove, "many people desire the discipline and control surveillance can bring."[57]

He notes that "[t]oo much social control, however, can adversely impact freedom, creativity, and self-development."[58] It would appear that potential risks to personal autonomy will depend, at least in part, on the pervasiveness of drone surveillance. Single, targeted operations would not appear to implicate this fear of societal control and harm to personal autonomy. However, some have questioned what effect 24-hour, *omnipresent* surveillance would have on our public spaces.[59] At this point, the shorter flight capabilities of small UAS—the aircraft likely to be most prevalent in the early stages of UAS integration—would limit the ability to conduct such pervasive surveillance. Yet, as this technology becomes more sophisticated and maximum possible flight times are extended, this risk to autonomy may heighten.

Anonymity

Anonymity is a "state of privacy" that "occurs when the individual is in public places or performing public acts but still seeks, and finds, freedom from identification and surveillance."[60] Professor Christopher Slobogin notes that "the right to public anonymity provides assurance that, when in public, one will remain nameless—unremarked, part of the undifferentiated crowd—as far as the government is concerned. The right is surrendered only when one does or says something that merits government attention, which most of the time must be something suggestive of criminal activity."[61] Others might argue that when we leave our doorstep, and enter into society, we give up this right to anonymity by permitting anyone to view our public movements and whereabouts. Potential blanket UAS surveillance over an urban landscape or a large public event (for instance, the Super Bowl or NASCAR racing), appears to pose a low risk to anonymity. Each person can remain a faceless person in the crowd, free from government identification. Take, for instance, CBP's surveillance at the border: its Predator-B drones reportedly fly at a minimum of 19,000 feet and are not able to identify specific persons on the ground.[62] However, the potential use of sensors such as Automated License Plate Readers (ALPRs)[63] or facial recognition technology,[64] or UAS surveillance specifically targeted at an individual, may have the capacity to eliminate one's anonymity when in public.

Post-Collection Activities

The second class of privacy risks posed by drone surveillance are those activities that occur after surveillance has been conducted—the aggregation, use, and retention of that data. In certain instances, the initial collection may not directly implicate an individual's privacy interests, but the subsequent manipulation and storage of that data may warrant an alternative privacy analysis

Aggregation

"Aggregation" is simply the gathering of information about a person whether from one or multiple sources.[65] In the pre-digital era, aggregating information about a person took concerted effort and lots of resources. With computer automation, information can be gathered and aggregated from many sources at the mere push of a button. The privacy theory of aggregation supposes that while the collection of bits of data about a person may not violate his or her privacy interests, extensive collection of information about him or her can rise to the level of a legal privacy intrusion. Professor Solove argues that this unique privacy intrusion arises because "when analyzed, aggregated information can reveal new facts about a person that she did not expect would be known about her when the original, isolated data was collected."[66] This theory was on display in the 2012 GPS tracking case *United States v. Jones*, where five Justices of the Supreme Court (in two separate concurrences) acknowledged that short-term location monitoring was likely not a Fourth Amendment search, but that 28 days of tracking should be considered a search.[67] Judge Richard Leon of the U.S. District Court for the

District of Columbia also applied this aggregation theory when invalidating the National Security Agency's (NSA's) metadata program, observing,

> the ubiquity of phones has dramatically altered the *quantity* of information that is now available and, *more importantly,* what that information can tell the Government about people's lives. ... Records that once would have revealed a few scattered tiles of information about a person now reveal an entire mosaic—a vibrant and constantly updating picture of the person's life.[68]

In the context of UAS operations, aggregation may mean the surveillance of an individual for an extended time, or the combination of drone-obtained data with other independent information. A simple one time fly-over of a residence may not significantly implicate any of the privacy interests described above (e.g., secrecy, autonomy, anonymity), but continuous surveillance of a person may implicate this aggregation interest. Alternatively, aggregating drone-collected data with other seemingly personal information, such as telephone,[69] banking,[70] utility,[71] and other records—all of which can be obtained without a probable cause warrant—might entail a unique privacy infringement beyond the mere collection of those individual data sets.

Use

The second post-collection privacy risk is the improper use of data—that is, data collected for an authorized purpose, but subsequently used in an unauthorized way. One commentator has argued that in the era of big data, where we inevitably share troves of personal information with the government and commercial entities, privacy rules that focus on the collection and retention of data are "becoming impractical for individuals, while also potentially cutting off future uses of data that could benefit society."[72] He argues that privacy laws should instead focus on controlling the *use* of that data.[73] Various federal laws embody this principle. For instance, the Privacy Act of 1974 requires any federal agency that maintains a database of personal records to inform each individual about whom it collects information of "the principal purpose or purposes for which the information is intended to be used."[74] Similarly, the Driver's Privacy Protection Act of 1994 made it a federal crime "for any person knowingly to obtain or disclose personal information, from a motor vehicle record, for any use not permitted" under that statute.[75] Instead of placing front-end restrictions on drone surveillance, such as requiring warrants before they are operated, policymakers might want to instead regulate how that information is used. For instance, a proposal may permit law enforcement officials to collect information for one purpose—say, traffic control—but prohibit it from using that information for other purposes, such as against an individual in a criminal prosecution absent a court order.

Retention

Another incident of the digital revolution is the near limitless ability of the government and private companies to store and retain information about individuals. The cost of storage has decreased exponentially over the past several decades and is no longer a hindrance to maintaining vast databases of personal data.[76] The Obama Administration's report on big data highlighted this concern when it noted that "the declining cost of collection, storage, and processing of data, combined with new sources of data like sensors, cameras, geospatial and other observational technologies, means that we live in a world of near-ubiquitous data collection."[77] This issue of data retention has been one of the drivers behind Europe's "right to be forgotten" law, which includes a provision requiring that data be retained "for no longer than is necessary for the purposes for which it was collected."[78] Generally speaking, when it comes to Fourth Amendment law, once information is lawfully collected, there are no additional constitutional hindrances to it being stored indefinitely. However, in the NSA metadata litigation, Judge Leon acknowledged that unique privacy interests were affected by the long-term storage of everyone's telephone call records.[79] He observed that in decades past, it was not expected that the government would retain a suspect's phone records once a case was concluded, whereas "[t]he NSA telephony metadata program ... involves the creation and maintenance of a historical database containing *five years'* worth of data."[80] The privacy impact of retention of UAS-derived data would largely depend on whether people could be identified in the recording. Retention of data that contains personally identifiable information and can be located based on this information would seem to implicate this retention interest.

VARIOUS APPROACHES TO UAS PRIVACY REGULATION

In addition to the general privacy concepts implicated by drone surveillance, one might question which government entity, if any, should regulate the potential privacy issues posed by the integration of thousands of UAS into domestic skies. Congress neither mentioned the word "privacy" in FMRA nor has it enacted any substantive privacy rules relating to drones subsequently. The FAA recently issued its proposed rule for operating small drones (55 pounds or less), but failed to include any privacy safeguards. Potentially in response to this dearth of privacy regulations, President Obama recently mandated that all federal agencies evaluate the privacy risks posed by their drone operations. This section will explore different approaches to UAS privacy regulation, focusing on the various government institutions—the courts, the executive branch, Congress, and state governments—that might conduct such regulation.

Courts

As the final arbiter of the Constitution,[81] the courts are frequently relied on to safeguard the privacy rights of Americans, whether through the suppression mechanism in criminal prosecutions,[82] or civil suits against government officials for violations of constitutional rights.[83] The Fourth Amendment, as applied by the courts, will provide a floor of legal protections against government actor use of drones. Likewise, privacy torts, which are given much of their content by the courts, will provide some legal recourse to potential privacy invasions caused by drones operated by private actors.

Fourth Amendment

The Fourth Amendment provides in relevant part: "The right of the people to be secure in their persons, houses, papers, and effects, against unreasonable searches and seizures, shall not be violated."[84] This Amendment, like most constitutional protections, applies only to acts by public actors,[85] and, as such, will provide the minimum legal requirements for government use of drones.[86] In order for the Fourth Amendment to apply in a particular situation, a reviewing court must assess whether the government conducted a "search" by asking whether it invaded an individual's "reasonable expectation of privacy."[87] Although no court has had the opportunity to apply the Fourth Amendment to drone technology, similar cases regarding traditional aircraft and location monitoring provide insight.

In three cases from the 1980s, the Supreme Court upheld the government's warrantless use of traditional aircraft to surveil both private and commercial property.[88] In *Florida v. Riley*, the case most closely resembling potential UAS surveillance, the police flew a helicopter 400 feet above a private residence to determine if marijuana was growing in a greenhouse in the backyard.[89] The Court held that this fly-over was not a Fourth Amendment search, as anyone from the public could have seen the property from that vantage point since the aircraft was in federal airspace.[90] Similarly, in the 1983 case *United States v. Knotts*, the Court held that it was not a Fourth Amendment search to track a person's public movements using an electronic tracking device.[91] There is a general consensus among commentators that a *strict* application of these cases would accord limited privacy safeguards to individuals located on both public and private property from UAS surveillance being conducted from lawful federal airspace.[92]

However, several more recent cases demonstrate that the Court has been uneasy about applying decade-old cases to new technology. In *Kyllo v. United States*, for instance, the use of sense-enhancing technology "not in general public use" to obtain information about the inside of a home was considered a Fourth Amendment search.[93] Similarly, in *United States v. Jones*, five Justices (in two separate concurrences) would have held that a month-long location monitoring using a GPS device constituted a search, even in the face of *Knotts*, which upheld the use of a more rudimentary tracking device.[94] The concurring Justices in *Jones* expressed concern about the sheer ability of these new technologies to collect vast amounts of information about individuals,[95] and their capacity to break down any natural barriers (such as time and resources) to excessive police surveillance.[96] This conundrum of applying old precedent to newer forms of technology is not unique for the federal courts. Several cases currently pending in the federal courts are wrestling with how to apply *Smith v. Maryland*, a case from the 1970s that addressed the collection of telephone numbers of one individual over several days,[97] to the NSA's collection of millions of telephone records over a five-year period.[98] With this trend in mind, federal judges may be persuaded that the sophistication and type of sensors used by UAS present a strong enough privacy risk to modify old precedents such as *Riley* or *Ciraolo* and adapt them to the new potential reality of UAS aerial surveillance. Additionally, depending on the duration of the surveillance and the amount of data collected on an individual, the courts may be inclined to apply an aggregation-type theory to long-term UAS surveillance.

This issue of the law keeping up with technology is a constantly recurring theme in Fourth Amendment jurisprudence. Some have argued that the judiciary is not the ideal forum for creating adequate privacy rules when fast-moving technology is involved.[99] Courts tend to be backward looking—resolving past factual scenarios between two discrete parties. This characteristic makes courts reactive rather than proactive, leading to privacy rules that might fall behind the particular technology in question. For instance, the Supreme

Court has yet to resolve whether individuals are entitled to a reasonable expectation of privacy in their emails, a technology that has been around for decades.[100] Part of the problem is that the Court has been unsure of its role in developing privacy rules when technology is in flux,[101] with some Justices preferring that legislatures, rather than the courts, take the lead role.[102] This is not to say that this approach is ineffective: this case-by-case approach allows the courts to formulate rules more cautiously based on concrete facts in an adversarial setting, and reduces the risk of creating rules with potentially unintended consequences.

This hesitancy to confront new technologies head-on might explain the Court's recent reliance on property rights as the basis for Fourth Amendment decisions, which, incidentally, might boost Fourth Amendment safeguards against UAS surveillance. In *Jones*, the five-Justice majority led by Justice Scalia rejuvenated a centuries old property-based theory of the Fourth Amendment by holding that the *physical attachment* of a GPS device on the undercarriage of a vehicle constituted an invasion of the owner's "effect" and therefore a Fourth Amendment search.[103] Justice Scalia observed that the *Katz* reasonable expectation of privacy formula added to, but did not replace, the traditional property-based test. Likewise, in *Florida v. Jardines*, the Court relied on this property theory in holding that the government's use of a trained police drug dog to investigate an individual's home and its immediate surroundings was a "search."[104] Depending on the flight path of the aircraft, and the height at which it is operating, this property theory might be employed to hold that flying a UAS onto someone's property with the intent to obtain information is a Fourth Amendment search for which a warrant would be required.

Privacy Torts

Like the Fourth Amendment, a body of laws collectively known as "privacy torts" might create safeguards against privacy invasions by both public- and private-actor use of unmanned aircraft. While some have held up privacy torts as proof that the existing body of case law is sufficient to regulate privacy issues arising from UAS operations,[105] others have asserted that such laws would provide minimal protection from drone surveillance, at least while in public.[106]

The genesis of the modern privacy tort sprung from the pens of Justice Louis Brandeis and Samuel Warren in their law review article "Right to Privacy."[107] There, they espoused the view that privacy could form the underpinning of civil liability absent other physical or tortious conduct. This new tort was later broken down into four discrete torts and included in a set of model rules intended for adoption by the states: (1) intrusion upon seclusion; (2) appropriation of one's name or likeness; (3) publicity given to private life; (4) publicity placing person in false light.[108] Of these four, the first and third will likely prove the most applicable to UAS surveillance.

The tort of intrusion upon seclusion, which may vary in its details from state to state, generally provides, "One who intentionally intrudes, physically or otherwise, upon the solitude or seclusion of another or his private affairs or concerns, is subject to liability to the other for invasion of his privacy, if the intrusion would be highly offensive to a reasonable person."[109] This tort applies a "reasonable person standard"—that is, it tests whether a person of "ordinary sensibilities" would be offended by the alleged invasion.[110] Likewise, the intrusion must not only be offensive, but "*highly* offensive,"[111] or as one court put it, "outrageously unreasonable conduct."[112] Generally, a single incident will not suffice; instead, the intrusion must be "repeated with such persistence and frequency as to amount to a course of hounding" and "becomes a burden to his existence.... "[113] However, in a few cases a single intrusion was adequate.[114] The invasion of privacy must be intentional, meaning the defendant must desire that the intrusion would occur, or, as with other torts,[115] know with a substantial certainty that such an invasion would result from his actions.[116] An accidental intrusion is not actionable. Finally, in some states, the intrusion must cause mental suffering, shame, or humiliation to permit recovery.[117]

The tort "publicity given to private facts" provides, "One who gives publicity to a matter concerning the private life of another is subject to liability to the other for invasion of his privacy, if the matter publicized is of a kind that (a) would be highly offensive to a reasonable person, and (b) is not of legitimate concern to the public."[118] Like the intrusion upon seclusion tort, publicity given to private facts focuses on publicity given to an individual's private, as opposed to public, life.[119] As the comments to the model rules observe, a person cannot complain of someone taking his photograph while walking down the street, but when a photograph is taken without his consent in a private place, and then subsequently published, he will have a valid publicity claim.[120] Unlike the intrusion tort, publicity given to private life requires that the information be made available to the public at large, and furthermore, generally prohibits claims that involve matters that are of legitimate public concern.[121]

Under current law, the location of the target of the surveillance largely controls whether someone has a viable claim for both intrusion upon seclusion and publicity given to private life, and this is likely to hold true with drone surveillance. For the most part, using a drone to peer inside the home of another–whether looking through a window or utilizing extra-sensory technology such as thermal imaging–would likely satisfy the intrusion tort, and if photographs were taken and subsequently published, that person would also likely have a claim for publicity given to private life. [122]

The likelihood of a successful claim is significantly diminished, however, if the surveillance is targeted at individuals in a public space,[123] or even while on private property so long as they could be viewed from a public vantage point.[124] Take, for instance, the suit brought by Aaron and Christine Boring against Google for photographs taken of their home and subsequently published online as part of Google's Street View program.[125] To create this program, Google employees attach panoramic cameras to their vehicles and drive around taking photographs of areas along the streets. The Borings, who live on a posted private road, discovered that Google had taken photographs of their private residence and swimming pool from their driveway. The Borings sued, among other claims, for intrusion upon seclusion and publicity given to private facts. Both claims were dismissed by the district court. On appeal, the Third Circuit Court of Appeals affirmed the dismissal on both counts, holding that the photographs failed to meet the "highly offensive" standard required under both privacy torts. The court found that "no person of ordinary sensibilities would be shamed, humiliated, or have suffered mentally as a result of a vehicle entering in his or her ungated driveway and photographing him or her from there."[126] The view of the Borings' house, pool, and driveway could "be seen by any person who entered onto their driveway, including a visitor or a delivery man."[127] Recording someone with a drone in a public place or even on private property that can be viewed from a public vantage point, as in the Google Street View case, would likely not constitute an invasion under either of these privacy torts.

Executive Branch

In addition to the courts, the executive branch is likely to play a significant role in regulating the privacy implications of UAS operations. Although the FAA has attempted to focus on safety and other regulatory issues, privacy remains a perennial topic for the FAA as it leads the federal government's UAS integration efforts. Some have posited that instead of the FAA, other executive branch agencies, such as DHS and DOJ, should take the lead role in overseeing the privacy implications of their own unmanned operations.[128]

President Obama recently issued a new memorandum taking a similar approach, directing all federal agencies to evaluate the privacy impact of their UAS operations and developing policies to mitigate such concerns.

FAA—Privacy Rules at the Test Sites and Beyond

In FRMA, Congress charged the FAA with integrating UAS into the national airspace.[129] Since passage of this legislation, the FAA has become the center of debate on the privacy implications of these new aircraft. Some in the industry and law enforcement have questioned FAA's authority to regulate privacy at all, arguing that it is both outside its expertise and its legislative grant of authority.[130] Others have argued that precisely because the FAA is charged with integrating UAS that it must resolve one of the most pressing issues involved with such integration—privacy.[131] As a general matter, the FAA has stated that its "mission does not include developing or enforcing policies pertaining to privacy or civil liberties,"[132] but several of its obligations in FMRA have kept the FAA enmeshed in this privacy debate.

The first such obligation was FMRA's mandate that the FAA establish six test sites to help facilitate the integration of UAS into the national airspace. Neither the program requirements nor the list of attributes for selecting the test sites expressly granted the FAA authority to regulate privacy, or even mentioned "privacy" as a general matter of consideration.[133] Nonetheless, the FAA sought public comment on a proposed privacy policy for UAS operations at the sites on February 22, 2013, and issued a final rule on November 14, 2013.[134] The privacy rules, which are contained as a provision in the contracts (known as an "Other Transaction Agreement," or OTA) between the FAA and each site operator include the following requirements: (1) operations at the test sites must adhere to existing federal, state, and local laws regarding an individual's right to privacy; (2) the test site operator must develop a publicly available privacy policy that is informed by the Fair Information Practice Principles (a generally accepted set of rules that regulate how data should be stored, used, and disseminated); (3) the site operator must maintain a record of all UAS operating at the test site; and (4) the site operator must require each UAS operator to have a written plan for the operator's use and retention of data collected by the UAS.[135] The authority to create these rules was purportedly under the FAA's contracting authority and not its general statutory authority.[136] Although the FAA retains the authority to rescind an OTA for a site operator in violation of existing privacy laws, ultimately, this privacy policy signals a hands-off approach, leaving the policing of privacy rules to private parties affected by operations at the test sites, to the test site operators themselves, and to local and state government bodies that oversee the test site operators.[137]

Despite this relatively passive approach to privacy at the test sites, the FAA has recognized the political and legal importance of privacy in both its proposed small UAS rule and its various planning documents required under FMRA. On February 15, 2015, the FAA issued its proposed operating requirements to allow small UAS (less than 55 pounds) to operate for non-hobby or non-recreational purposes.[138] As was expected, the FAA noted that privacy concerns "were beyond the scope of this rulemaking."[139] However, the FAA also noted it would participate in the "multi-stakeholder engagement process" (described below) to assist in a privacy framework concerning commercial and private use of drones. Also, in its five-year roadmap required under FMRA, the FAA noted that while its primary mission "does not include developing or enforcing policies pertaining to privacy or civil liberties, experience with the UAS test sites will present an opportunity to inform the dialogue ... concerning the use of UAS technologies and the areas of privacy and civil liberties."[140] Likewise, in its Comprehensive Plan, also required under FMRA, the FAA devoted a whole section to highlighting the privacy and civil liberties concerns that all federal agencies must take into account as UAS are integrated into the national airspace.[141] While safety will undoubtedly remain the top priority of FAA officials as it navigates the difficult task of integrating drones in the national airspace, with its prominent role of testing and licensing both government, commercial, and private use of drones, it will remain a significant voice in the ongoing privacy debate.

President's Memorandum on UAS and Privacy

On February 15, 2015, the same day the FAA issued its proposed small UAS rule, President Obama issued a memorandum entitled "Promoting Economic Competitiveness While Safeguarding Privacy, Civil Rights, and Civil Liberties in Domestic Use of Unmanned Aircraft Systems."[142] This presidential memorandum establishes two new frameworks relating to the potential privacy implications of drone operations by both government and private actors.

UAS Policies and Procedures for Federal Government Use

The memorandum first observes that all operations conducted by federal agencies must comply with the Constitution, federal law, and other applicable regulations and policies, an obligation to which agencies are already subject. The memorandum requires that, prior to deployment of new UAS technology and at least every three years, all federal agencies must "examine their existing UAS policies and procedures relating to the collection, use, retention, and dissemination of information obtained by UAS, to ensure that privacy, civil rights, and civil liberties are protected."[143] Agencies that collect information through UAS shall ensure their policies and procedures adhere to the following requirements:

i. Collection and Use. Agencies must only collect information using UAS, or UAS-collected information, to the extent that such collection or use is consistent with and relevant to an authorized purpose.
ii. Retention. Information collected using UAS that may contain PII [personally identifiable information] shall not be retained for more than 180 days unless retention of the information is determined to be necessary to an authorized mission of the retaining agency, is maintained in a system of records covered by the Privacy Act, or is required to be retained for a longer period by any other applicable law or regulation.
iii. Dissemination. UAS-collected information that is not maintained in a system of records covered by the Privacy Act shall not be disseminated outside of the agency unless dissemination is required by law, or fulfills an authorized purpose and complies with agency requirements.

To ensure accountability, the memorandum requires, among other things, that agencies develop protocols for receiving, investigating, and addressing potential privacy complaints; ensure they have rules of conduct and training for federal government personnel and contractors; establish meaningful oversight of individuals who have access to sensitive information collected by UAS; develop policies and procedures to authorize the use of UAS in response to a request for UAS assistance of federal, state, tribal, or territorial government operations; and ensure that local entities that purchase UAS through federal funding have policies in place to safeguard privacy and civil liberties prior to expending such funds.

To promote transparency, the memorandum requires agencies to provide notice to the public regarding where the agency's UAS are authorized to operate; keep the public informed about the agency's UAS program as well as changes that would significantly affect privacy; and make available to the public, on an annual basis, a general summary of the agency's UAS operations during the previous fiscal year, to include a brief description of types or categories of missions flown, and the number of times the agency provided assistance to other agencies, or to state, local, tribal, or territorial governments.

The agencies must report to the President within 180 days from the date of the issuance of the memorandum with a status on the implementation of these policies and procedures, and must make these policies publicly available within one year.

Multi-Stakeholder Engagement Process for Commercial and Private UAS Use

The President's memorandum also charges the Department of Commerce, through the National Telecommunications and Information Administration (NTIA), to initiate "a multi-stakeholder engagement process to develop a privacy framework regarding privacy, accountability, and transparency for commercial and private UAS use."[144] The end result is expected to be a set of voluntary best practices for privacy issues implicated by commercial and private UAS use.

Privacy Impact Assessments and Privacy Working Groups

Prior to the President's memorandum, several federal agencies had taken steps to assess and address the privacy impacts of their UAS operations. In September 2013, DHS's Chief Privacy Officer, in conjunction with CBP's Office of Air and Marine, issued a Privacy Impact Assessment (PIA) primarily evaluating the privacy impact of its operations at the U.S. border.[145] This was prompted, in part, by statutory obligations under federal law.

Under the E-Government Act of 2002, federal agencies must conduct a PIA before (1) "developing or procuring information technology that collects, maintains, or disseminates information that is in an identifiable form"; or (2) "initiating a new collection of information that will be collected, maintained, or disseminated using information technology ... [and] includes information in an identifiable form permitting the physical or online contacting of a specific individual."[146] Specific to the Department of Homeland Security, under the Homeland Security Act of 2002, DHS must conduct PIAs "of proposed rules of the Department or that of the Department on the privacy of personal information, including the type of personal information collected and the number of people affected."[147] More generally, the Privacy Officer of DHS is required to coordinate with the Officer for Civil Rights and Civil Liberties to ensure that "(A) programs, policies, and procedures involving civil rights, civil liberties, and privacy considerations are addressed in an integrated and comprehensive manner; and (B) Congress receives appropriate reports on such programs, policies, and procedures."[148]

DHS's PIA evaluated the various uses of UAS in DHS's operations, including its operations at the border, and concluded that such operations have minimal privacy impact. Because current DHS operations limit flights to an altitude of 19,000 feet, the report noted that cameras on UAS do not permit the operator to specifically identify people on the ground nor do they use technology that sees through walls or collects information regarding the interior of buildings. The report noted that personally identifiable information (PII) is generally not retrieved, but that images may later be associated with specific individuals if they are apprehended and the surveillance feed is associated with them. Acknowledging that UAS can stay in flight longer than traditional aircraft, the report concludes that the privacy impact of this technological difference is mitigated by various safeguards including well-defined flight operations, restrictions on accessing collected data, and security protocols for storing data. It is likely that DHS will utilize this PIA when developing any new policies and procedures required by the President's new memorandum.

In addition to PIAs, in late 2012, DHS's Office for Civil Rights and Civil Liberties and its Privacy Office spearheaded the "Working Group to Safeguard Privacy, Civil Rights, and Civil Liberties in the Department's Use and Support of Unmanned Aerial Systems."[149] The working group was intended to assess all active and planned uses of UAS by the various components of DHS and flag any privacy or civil liberties issues potentially raised by such operations. The goal was to develop a set of best practices for safeguarding these various legal interests, and it was reported in September 2014 that such a document would be out by the end of 2014.[150] DHS has yet to release this best practices document. It may be that this document will be subsumed into the larger obligations placed on DHS through the President's new memorandum.

DOJ Inspector General UAS Report

There has been some debate within DOJ concerning the extent to which unmanned aerial surveillance differs from manned aerial surveillance, and whether these differences warrant new rules for deploying UAS in law enforcement operations. A report issued by DOJ's Office of Inspector General explained that both the FBI and ATF did not see a practical difference in how UAS collect evidence as compared to their manned counterparts.[151] As such, these entities did not see a need to develop specialized UAS privacy protocols.

DOJ's Inspector General disagreed with the assessment of the FBI and ATF and concluded the following in its interim report:

> We found that the technological capabilities of UAS and the current, uncoordinated approach of DOJ components to UAS use may merit the DOJ developing consistent, UAS-specific policies to guide the proper use of UAS. Unlike manned aircraft, UAS can be used in close proximity to a home and, with longer-lasting power systems, may be capable of flying for several hours or even days at a time, raising unique concerns about privacy and the collection of evidence with UAS. Considering that multiple components are using or have the potential to use UAS, we believe the Office of the Deputy Attorney General (ODAG), which has the primary responsibility within DOJ for formulating cross-component law enforcement policies, should consider the need for a DOJ-wide policy regarding UAS uses that could have significant privacy or other legal implications.[152]

This conclusion that unmanned operations pose a unique and potentially greater privacy threat than manned operations mirrors that of the President's new memorandum on UAS operations. Even when DOJ, like all other federal agencies using UAS, adopts specific policies and procedures addressing the privacy impact of its UAS operations, this inevitably prompts the question whether federal agencies should be left to police their own surveillance activities. While praising the White House drone memorandum as "an important and welcome step in advancing drone technology," one commentator noted that the memo itself "does not establish strong privacy and transparency drone standards for agencies, leaving it up to the agencies to develop these standards."[153] He continues: "Because the memo's requirements are not specific, the drone policies the agencies set for themselves will be key to how individuals' privacy is actually protected. Congress still has a role to play in setting strong privacy and transparency standards for drone use."[154]

Congress

Several measures were introduced in the 113th Congress that would have restricted both public-and private-actor domestic UAS operations, and reintroduction of these bills is likely in the 114th Congress. The proposed regulations in these bills range from warrant requirements for law enforcement operations to comprehensive data collection and minimization statements to privacytort-like prohibitions against private-actor intrusions. Additionally, Congress has utilized its oversight authority to hold hearings and probe executive branch agencies to disclose when and where they are using drones and the potential privacy implications of such uses.

Legislation

Drone Aircraft Privacy and Transparency Act of 2013 (S. 1639, H.R. 2868)

In the 113th Congress, Senator Ed Markey and Representative Peter Welch introduced nearly identical legislation entitled the Drone Aircraft Privacy and Transparency Act of 2013 (S. 1639, H.R. 2868).[155] These bills would have amended FMRA to create a comprehensive scheme to regulate both government and private-actor use of drones, including data collection requirements, a warrant requirement for law enforcement, and various enforcement mechanisms.

First, these bills would have required the Secretary of Transportation, with input from the Secretary of Commerce, the Chairman of the Federal Trade Commission, and the Chief Privacy Officer of the Department of Homeland Security, to study any potential threats to privacy protections posed by the introduction of drones in the national airspace. They would have prohibited the FAA from issuing a license to operate a drone unless the application included a "data collection statement." This statement would have had to include a list of individuals who would have the authority to operate the drone; the location in which the drone would be used; the maximum period it would be used; and whether the drone would be collecting information about individuals. If the drone would be used to collect personal information, the statement would have had to provide the circumstances in which such information would be used; the kinds of information collected and the conclusions drawn from it; the type of data minimization procedures to be employed; whether the information would be sold, and if so, under what circumstances; how long the information would be stored; and the procedures for destroying irrelevant data.

The statement also would have had to include information about the possible impact on privacy protections posed by the operation under that license and steps to be taken to mitigate this impact. Additionally, the statement would have had to include the contact information of the drone operator; a process for determining what information had been collected about an individual; and a process for challenging the accuracy of such data. Finally, the FAA would have been required to post the data collection statement on the Internet.

In addition to the data collection statement, any law enforcement agency which operates a drone would have had to file a "data minimization statement" with the FAA. This statement would have required adoption of policies by the agency that minimize the collection of information and data unrelated to the investigation of a crime under a warrant; required the destruction of data that is no longer relevant to the investigation of a crime; established procedures for the method of such destruction; and established oversight and audit procedures to ensure the agency operates a UAS in accordance with the data collection statement filed with the FAA.

S. 1639 and H.R. 2868 would have provided several enforcement mechanisms. First, the FAA could have revoked a license of a user that did not comply with these requirements. The Federal Trade Commission would have had the primary authority to enforce the data collection requirements. Additionally, the Attorney General of each state, or an official or agency of a state, would have been empowered to file a civil suit if there was reason to believe that the privacy interests of residents of that state had been threatened or adversely affected. These bills would have also created a private right of action for a person injured by a violation of this legislation.

These bills would also have prohibited a governmental entity from using a drone, or obtaining information from another person using a drone, for protective activities, or for law enforcement or intelligence purposes, except with a warrant. This prohibition would not apply in "exigent circumstances," which was defined to mean imminent danger of death or serious physical injury or high risk of terrorist attack as determined by the Secretary of Homeland Security.

Preserving Freedom from Unwarranted Surveillance Act of 2013 (S. 1016, H.R. 972)

Senator Rand Paul and Representative Austin Scott's companion bills, the Preserving Freedom from Unwarranted Surveillance Act of 2013 (S. 1016, H.R. 972), would have focused exclusively on government drone operations.[156] These bills would have required any entity acting under the authority of the federal government to obtain a warrant based upon probable cause before conducting drone surveillance to investigate violations of criminal law or regulations. S. 1016 and H.R. 972 included several exceptions to this warrant requirement: (1) when necessary to prevent or deter illegal entry of any persons or illegal substances into the United States; (2) when a law enforcement officer possesses reasonable suspicion that under particular circumstances "swift action to prevent imminent danger to life or serious damage to property, or to forestall the imminent escape of a suspect, or destruction of evidence" is necessary; or (3) when the Secretary of Homeland Security determines credible intelligence indicates a high risk of a terrorist attack by a specific individual or organization. H.R. 972 would have created a right to sue for any violation of its prohibitions. Unlike H.R. 972, S. 1016 included an express exclusionary rule for evidence obtained in violation of the act.

Preserving American Privacy Act of 2013 (H.R. 637)

Representative Ted Poe introduced the Preserving American Privacy Act of 2013 (H.R. 637), which would have regulated both public and private use of drones under various mechanisms.[157] As to law enforcement use, H.R. 637 would have created a general prohibition on the use of drones to collect covered information or disclose covered information so collected. "Covered information" was defined as "information that is reasonably likely to enable identification of an individual" or "information about an individual's property that is not in plain view." This prohibition was subject to the following exceptions: (1) law enforcement obtains a court-issued warrant and serves a copy of the warrant on the target of the search within 10 days of the surveillance. However, notice need not be provided if it would jeopardize an ongoing criminal or national security investigation.(2) Law enforcement obtains a court-issued order based upon "specific and articulable facts showing a reasonable suspicion of criminal activity and a reasonable probability" that the operation "will provide evidence of such criminal activity." The order may authorize surveillance in a stipulated public area for no more than 48 hours which may be renewed for a total of 30 days. Notice of the operation must be provided to the target no later than 10 days after the operation. Alternatively, notice may be provided not less than 48 hours before the operation in a major publication, on a government website, or with signs posted in the area of the operation. (3) Operation is within 25 miles of national border. (4)The targeted individual has provided prior written consent. (5) Emergency situation involves danger of death or serious physical injury, conspiratorial activities threatening the national security interest, or conspiratorial activities characteristic of organized crime, where a warrant cannot be obtained with due diligence. Law enforcement must then obtain a warrant within 48 hours of such operation. Any evidence obtained in violation of this act would not have been admissible in any trial or adjudicative proceeding.

Additionally, H.R. 637 would have required any governmental entity applying for a certificate or license to operate a UAS to also file a data collection statement with the Attorney General, which would have included the purpose for which the UAS will be used; whether the UAS is capable of collecting covered information; the length of time the information will be retained; a point of contact for citizen feedback; the particular unit of governmental entity responsible for safe and appropriate operation of the UAS; the rank and title of the individual who may authorize the operation of the UAS; the applicable data minimization policies barring the collection of covered information unrelated to the investigation of crime and requiring the destruction of covered information that is no longer relevant to the investigation of a crime; and applicable audit and oversight procedures.

Under H.R. 637, the Attorney General would have been empowered to request that the Secretary of Transportation revoke the license or certificate of any entity that fails to file a data collection statement. Further, H.R. 637 contained a provision permitting administrative discipline against an officer who intentionally violates a provision of this act.

H.R. 637 would also have made it unlawful to intentionally operate a private UAS to capture images in a manner highly offensive to a reasonable person where the person is engaging in a personal or familial activity under circumstances in which the individual has a reasonable expectation of privacy, regardless of whether there is a physical trespass.[158]

Oversight

In addition to its authority to enact federal law, Congress can and has utilized its oversight function to shape the debate surrounding the privacy implications of drone surveillance. For instance, both the Senate and House Judiciary Committees held hearings in the 113th Congress specifically addressing the privacy impact of drone operations. Additionally, individual members probed executive branch officials on when and where they were using drones and the potential privacy implications of such uses. One such example of this oversight function is demonstrated by correspondence between the Senator Rand Paul and the FBI in the summer of 2013. In a July 25, 2013, letter to FBI Director Mueller, Senator Paul asked how the Bureau defined "reasonable expectation of privacy" in the context of UAS surveillance, as Paul feared that "an overbroad interpretation of this protection would enable more substantial information collection on an individual in a circumstance they might not have believed was subject to surveillance."[159] The FBI responded by arguing that based on the Supreme Court's three aerial surveillance cases from the 1980s, it need not obtain a warrant for any surveillance "open to public view" and that the "Fourth Amendment principles applicable to manned aerial surveillance discussed in these cases apply equally to UAVs."[160] Similar oversight is likely to continue in the 114th Congress.

States

Justice Brandeis once observed that "it is one of the happy incidents of the federal system that a single courageous state may, if its citizens choose, serve as a laboratory; and try novel social and economic experiments without risk to the rest of the country."[161] Some have argued that under this theory of federalism, the states are best situated to experiment with various UAS policies to determine the most appropriate legal framework for their state.[162] Others have countered that applying a patchwork of 50 different privacy regimes would be overburdensome on both public and private users of UAS. According to the National Conference of State Legislatures (NCSL), by the end of 2014, 20 states had enacted laws addressing UAS issues.[163]

State laws in this area have mainly come in three forms. The first, and perhaps most numerous, are laws that create broad prohibitions on their own state law enforcement entities, subject to various exceptions that differ from state to state.[164] These exceptions include obtaining a warrant based upon probable cause; countering a high risk of terrorist attack; responding to threats of imminent harm of human life; locating a missing person; and conducting crime scene and traffic photography, among others. The second category relates to use restrictions of data collected by UAS. For instance, Illinois requires law enforcement to destroy all information gathered by a UAS within 30 days of collection unless there is reasonable suspicion that the information contains evidence of criminal activity or the information is relevant to an ongoing criminal investigation.[165] While the relevancy standard is a very low evidentiary threshold, it does prevent limitless retention of private data. Similarly, Alaska prohibits the retention of records collected by drones unless it is required as part of an investigation or prosecution, is used for training purposes, or is required by federal or state law.[166] Images that are retained under Alaska's statute are considered confidential and not subject to the state's public records laws. The third category relates to regulation of private actors, generally through the creation of private causes of action for privacy invasions caused by UAS surveillance. For instance, Idaho prohibits recording individuals on their residence without their consent, or from photographing an individual for the purpose of publishing or otherwise publicly disseminating such photograph no matter where the target is located.[167] Undoubtedly, this third category of proposals creates tension between the public's First Amendment right to gather news and the individual privacy interests at stake, requiring legislatures to fine tune this balance when enacting UAS legislation.

End Notes

[1] FAA Modernization and Reform Act of 2012, P.L. 112-95, §§331-336, 126 Stat. 11, 72-78.

[2] *See* Operation and Certification of Small Unmanned Aircraft Systems, 80 Fed. Reg. 9544, 9552 (proposed Feb. 23, 2015).

[3] *See, e.g.*, *The Future of Drones In America: Law Enforcement and Privacy Considerations: Hearing Before the S. Comm. on the Judiciary*, 113th Cong. (2013) (hereinafter Senate Judiciary Hearing); *Eyes in the Sky: The Domestic Use of Unmanned Aerial Systems: Hearing Before the H. Comm. on the Judiciary Subcomm. on Crime, Terrorism, Homeland Security, and Investigations*, 113th Cong. (2013).

[4] *See* Drone Aircraft Privacy and Transparency Act of 2013, S. 1639, 113th Cong. (2013); H.R. 2868, 113th Cong. (2013); Preserving Freedom from Unwarranted Surveillance Act of 2013, H.R. 972, 113th Cong. (2013), S. 3287, 112th Cong. (2012); Preserving American Privacy Act of 2013, H.R. 637, 113th Cong. (2013).

[5] *See* "Executive Branch," *infra* p. 17.

[6] *See* "States," *infra* p. 26.

[7] *See, e.g.*, California v. Ciraolo, 476 U.S. 207, 215 (1986).

[8] PRESIDENTIAL MEMORANDUM: PROMOTING ECONOMIC COMPETITIVENESS WHILE SAFEGUARDING PRIVACY, CIVIL RIGHTS, AND CIVIL LIBERTIES IN DOMESTIC USE OF UNMANNED AIRCRAFT SYSTEMS (Feb. 15, 2015) [hereinafter Presidential Memorandum on UAS].

[9] Operation and Certification of Small Unmanned Aircraft Systems, 80 Fed. Reg. at 9552.

[10] U.S. CONST. art. I, §8, cl. 3.

[11] *See* Drone Aircraft Privacy and Transparency Act of 2013, S. 1639, 113th Cong. (2013); H.R. 2868, 113th Cong. (2013); Preserving Freedom from Unwarranted Surveillance Act of 2013, H.R. 972, 113th Cong. (2013), S. 3287, 112th Cong. (2012); Preserving American Privacy Act of 2013, H.R. 637, 113th Cong. (2013).

[12] *See* WELLS C. BENNETT, CIVILIAN DRONES, PRIVACY, AND THE FEDERAL-STATE BALANCE, BROOKINGS INSTITUTION (2014); Margot E. Kaminski, *Drone Federalism, Civilian Drones and the Things They Carry*, 4 CAL. L. REV. CIRCUIT 57 (2013).

[13] National Conference of State Legislatures, 2014 Unmanned Aircraft Systems (UAS) Legislation, *available at* http://www.ncsl.org/research/civil-and-criminal-justice/2014-state-unmanned-aircraft-systems-uas-legislation.aspx

[14] For a more in-depth look at privacy and drones, see CRS Report R42701, *Drones in Domestic Surveillance Operations: Fourth Amendment Implications and Legislative Responses*, by

Richard M. Thompson II and CRS Report R42940, *Integration of Drones into Domestic Airspace: Selected Legal Issues*, by Alissa M. Dolan and Richard M. Thompson II.

[15] P.L. 112-95, §331, 126 Stat. 72.

[16] *Id.*

[17] See CRS Report R42136, *U.S. Unmanned Aerial Systems*, by Jeremiah Gertler, for a description of the various types of drones operated in the United States.

[18] *US Army unveils 1.8 gigapixel camera helicopter drone*, BBC NEWS (December 29, 2011 6:11 p.m.), http://www.bbc.com/news/technology-16358851.

[19] *See* Draganflyer X6, Thermal Infrared Camera, http://www.draganfly.com/uav-helicopter/draganflyer-x6/features/ flir-camera.php.

[20] This sensor can recognize and permit drones to track vehicles based on license plate numbers. Customs and Border Protection Today, Unmanned Aerial Vehicles Support Border Security (July 2004), http://www.cbp.gov/xp/ CustomsToday/2004/Aug/other/aerial_vehicles.xml.

[21] This sensor produces three-dimensional images, and has the capability to see through trees and foliage. U.S. ARMY, UAS CENTER FOR EXCELLENCE, "EYES OF THE ARMY" US ARMY ROADMAP FOR UNMANNED AIRCRAFT SYSTEMS 2010- 2035, at 83 (2010).

[22] *See* Clay Dillow, *Army Developing Drones that Can Recognize Your Face from a Distance*, POPSCI (September 28, 2011, 5:01 p.m.), *available at* http://www.popsci.com/ technology/article/2011-09/army-wants-drones-can-recognizeyour-face-and-read-your-mind. The issue of weaponizing drones, which presents unique legal issues, is beyond the scope of this report.

[23] Fast Lightweight Autonomy (FLA), Defense Sciences Office, Defense Advanced Research Projects Agency (last visited Feb. 24, 2015), http://www.darpa.mil/Our_Work/ DSO/ Programs/Fast_Lightweight_Autonomy_%28FLA%29.aspx.

[24] FAA, "Unmanned Aircraft Operations in the National Airspace System," 72 Fed. Reg. 6689 (Feb. 13, 2007).

[25] *Id.*

[26] *See generally* FAA "Unmanned Aircraft Systems," *available at* http://www.faa.gov/ about/initiatives/uas/cert/.

[27] Freedom of Information Act Responses, FEDERAL AVIATION ADMINISTRATION (last visited March 4, 2015), https://www.faa.gov/uas/public_operations/foia_responses/.

[28] Before operating a UAS in the national airspace, CBP, as with any other governmental entity, must obtain a Certificate of Authorization (COA) from the FAA. See CRS Report R42940, for a discussion of FAA's UAS licensing process.

[29] DEP'T OF HOMELAND SECURITY, PRIVACY IMPACT ASSESSMENT FOR THE AIRCRAFT SYSTEMS 4 (2013).

[30] Customs and Border Protection Drone Flight List – Supplemental, Electronic Frontier Foundation (last visited Feb. 19, 2015), *available at* https://www.eff.org/document/eff-v-dhs-cbp-supplemental-agency-list.

[31] *See* Jason Koebler, *First Man Arrested With Drone Evidence Vows to Fight Case*, U.S. NEWS (April 9, 2012), *available at* http://www.usnews.com/news/articles/2012/04/09/first-man-arrested-with-drone-evidence-vows-to-fightcase.

[32] Customs and Border Protection Drone Flight List, *supra* note 30.

[33] U.S. Dep't of Justice, Office of the Inspector General, Interim Report on the Department of Justice's Use and Support of Unmanned Aircraft Systems (2013), *available at* http://www.justice.gov/oig/reports/2013/a1337.pdf.

[34] *Id.* at 5.

[35] Letter from Stephen D. Kelly, Office of Congressional Affairs, Federal Bureau of Investigation to Senator Rand Paul (July 19, 2013).

[36] DOJ Inspector General Report, *supra* note 33, at 5.

[37] *Id.* at 6.

[38] *See* 2012 FAA Response to Congressman Markey on Drones, Electronic Frontier Foundation (Sept. 21, 2012), *available at* https://www.eff.org/document/2012-faa-response-cong ressman-markey-drones.

[39] *Id.*

[40] *Id.*

[41] *See* 72 Fed. Reg. 6689; 14 C.F.R. §§21.191, 21.193 (experimental certificates generally); 14 C.F.R. §91.319 (operating limitations on experimental certificate aircraft).

[42] P.L. 112-95, §333, 126 Stat. 75. (July 19, 2013).

[36] DOJ Inspector General Report, *supra* note 33, at 5.

[37] *Id.* at 6.

[38] *See* 2012 FAA Response to Congressman Markey on Drones, Electronic Frontier Foundation (Sept. 21, 2012), *available at* https://www.eff.org/document/2012-faa-response-congres sman-markey-drones.

[39] *Id.*

[40] *Id.*

[41] *See* 72 Fed. Reg. 6689; 14 C.F.R. §§21.191, 21.193 (experimental certificates generally); 14 C.F.R. §91.319 (operating limitations on experimental certificate aircraft).

[42] P.L. 112-95, §333, 126 Stat. 75.

[43] Authorizations Granted Via Section 333 Exemptions, Federal Aviation Administration (last visited Feb. 19, 2015), *available at* https://www.faa.gov/uas/legislative_programs/ section_ 333/333_authorizations/.

[44] P.L. 112-95, §336, 126 Stat. 77.

[45] *See* Sierra Nevada Corporation Achieves Milestone for USAF's Advanced Wide-Area Airborne Persistent Surveillance (WAPS) System – Gorgon Stare Increment 2 (July 14, 2014), *available at* http://www.sncorp.com/ AboutUs/NewsDetails/618.

[46] *See Sierra Nevada fields Argus-IS upgrade to Gorgon Stare pod, Flight Global* (July 2, 2014), *available at* http://www.flightglobal.com/news/articles/sierra-nevada-fields-argus-is-up grade -to-gorgon-stare-400978/.

[47] Privacy Impact Assessment, *supra* note 29, at 8.

[48] *See* Victor Blackwell & Michael Pearson, *FBI: Bombs Found in Alabama Kidnapper's Bunker*, CNN (Feb. 5, 2013), *available at* http://www.cnn.com/2013/02/05/us/alabama-child-hostage/.

[49] ALAN WESTIN, PRIVACY AND FREEDOM 7 (1967).

[50] *See* Daniel J. Solove, *A Taxonomy of Privacy*, 154 U. PA. L. REV. 477, 494 (2006).

[51] *See* Florida v. Riley, 488 U.S. 445, 450-51 (1989); Dow Chemical Co. v. United States, 476 U.S. 227, 239 (1986); California v. Ciraolo, 476 U.S. 207, 215 (1986).

[52] *Ciraolo*, 476 U.S. at 213 (1986) (quoting Katz v. United States, 389 U.S. 347, 351 (1967)).

[53] See "Privacy Torts," *infra* p. 14.

[54] *See* Whalen v. Roe, 429 U.S. 589, 599 (1977).

[55] *See* JOHN GILLIOM, OVERSEERS OF THE POOR: SURVEILLANCE, RESISTANCE, AND THE LIMITS OF PRIVACY 3 (2001) ("Surveillance of human behavior is in place to control human behavior, whether by limiting access to programs or institutions, monitoring and affecting behavior within those arenas, or otherwise enforcing rules and norms by observing and recording acts of compliance and deviance."); Julie E. Cohen, *Examined*

Lives: Informational Privacy and the Subject as Object, 52 STAN. L. REV. 1373, 1426 (2000) ("The point is not that people will not learn under conditions of no-privacy, but that they will learn differently, and that the experience of being watched will constrain, ex ante, the acceptable spectrum of belief and behavior. Pervasive monitoring of every first move or false start will, at the margin, incline choices toward the bland and the mainstream. The result will be a subtle yet fundamental shift in the content of our character, a blunting and blurring of rough edges and sharp lines.").

[56] 4 JEREMY BENTHAM, THE WORKS OF JEREMY BENTHAM 40 (1843).

[57] Solove, *supra* note 50, at 494.

[58] *Id.*

[59] *See* JAY STANLEY AND CATHERINE CRUMP, PROTECTING PRIVACY FROM AERIAL SURVEILLANCE: RECOMMENDATIONS FOR GOVERNMENT USE OF DRONE AIRCRAFT, AMERICAN CIVIL LIBERTIES UNION 11 (2011).

[60] Westin, *supra* note 49, at 7.

[61] Christopher Slobogin, *Public Privacy: Camera Surveillance of Public Places and the Right to Anonymity*, 72 MISS. L.J. 213, 238 (2002).

[62] Privacy Impact Assessment, *supra* note 29, at 7.

[63] Customs and Border Protection Today, Unmanned Aerial Vehicles Support Border Security (July 2004), http://www.cbp.gov/xp/CustomsToday/2004/Aug/other/aerial_vehicles.xml.

[64] *See* Andrew Conte, *Drones with Facial Recognition Technology Will End Anonymity, Everywhere*, Business Insider (May 27, 2013), *available at* http://www.businessinsider.com/facial-recognition-technology-and-drones-2013-5.

[65] *See* Solove, *supra* note 50, at 507.

[66] *Id.*

[67] United States v. Jones, 132 S. Ct. 945, 954 (2012) (Sotomayor, J., concurring), *id.* at 957 (Alito, J., concurring in the judgment).

[68] Klayman v. Obama, 957 F. Supp. 2d 1, 35-36 (D.D.C. 2013). This case is currently before the D.C. Circuit Court of Appeals.

[69] *See* Smith v. Maryland, 442 U.S. 735 (1979).

[70] *See* United States v. Miller, 425 U.S. 435 (1976).

[71] *See* United States v. McIntyre, 646 F.3d 1107 (8th Cir. 2011); United States v. Starkweather, No. 91-30354, 1992 WL 204005, at *2 (9th Cir. August 24, 1992).

[72] *See* Craig Mundie, *Privacy Pragmatism*, FOREIGN AFFAIRS (April 2014), *available at* http://www.foreignaffairs.com/ articles/140741/craig-mundie/privacy-pragmatism.

[73] *Id.*

[74] 5 U.S.C. §552a(e)(3)(B).

[75] 18 U.S.C. §2722(a).

[76] *See generally* Orin S. Kerr, *The Next Generation Communications Privacy Act*, 162 U. PA. L. REV. 373, 391 (2014).

[77] EXECUTIVE OFFICE OF THE PRESIDENT, BIG DATA: SEIZING OPPORTUNITIES, PRESERVING VALUES (2014), *available at* http://www.whitehouse.gov/sites/default/files/docs/big_data_privacy_report_5.1.14_final_print.pdf.

[78] *See* Directive 95/46/EC of the European Parliament and the Council of 24 October 1995 on the Protection of Individuals with Regard to Processing of Personal Data and on the Free Movement of Such Data, 1995 O.J. (L 281) Art. 6.

[79] *Klayman*, 976 F. Supp. 2d at 32.

[80] *Id.* (emphasis in original).

[81] *See* Marbury v. Madison, 5 U.S. 137, 177 (1803) ("It is emphatically the province and duty of the judicial department to say what the law is."); Cooper v. Aaron, 358 U.S. 1, 18 (1958) ("Marbury declared the basic principle that the federal judiciary is supreme in the exposition of the law of the Constitution, and that principle has ever since been respected by this Court and the Country as a permanent and indispensable feature of our constitutional system.").

[82] *See* Mapp v. Ohio, 367 U.S. 643 (1961).

[83] *See* 42 U.S.C. §1983.

[84] U.S. CONST. amend. IV.

[85] Burdeau v. McDowell, 256 U.S. 465, 475 (1921) ("The Fourth Amendment gives protection against unlawful searches and seizures, and as shown in the previous cases, its protection applies to governmental action. Its origin and history clearly show that it was intended as a restraint upon the activities of sovereign authority, and was not intended to be a limitation upon other than governmental agencies").

[86] For a more detailed look at how the Fourth Amendment might apply to UAS, see CRS Report R42701, *Drones in Domestic Surveillance Operations: Fourth Amendment Implications and Legislative Responses*, by Richard M. Thompson II.

[87] *See* Katz v. United States, 389 U.S. 347, 362 (Harlan, J., concurring).

[88] *See* Florida v. Riley, 488 U.S. 445, 450-51 (1989) ("Riley could not reasonably have expected that his greenhouse was protected from public or official observation from a helicopter had it been flying within the navigable airspace for fixed-wing aircraft."); Dow Chemical Co. v. United States, 476 U.S. 227, 239 (1986) ("We hold that the taking of aerial photographs of an industrial plant complex from navigable airspace is not a search prohibited by the Fourth Amendment."); California v. Ciraolo, 476 U.S. 207, 215 (1986) ("In an age where private and commercial flight in the public airways is routine, it is unreasonable for respondent to expect that his marijuana plants were constitutionally protected from being observed with the naked eye from an altitude of 1,000 feet. The Fourth Amendment simply does not require the police traveling in the public airways at this altitude to obtain a warrant in order to observe what is visible to the naked eye.").

[89] *Riley*, 488 U.S.at 448.

[90] *Ciraolo*, 476 U.S. at 213 ("The Fourth Amendment protection of the home has never been extended to require law enforcement officers to shield their eyes when passing by a home on public thoroughfares. Nor does the mere fact that an individual has taken measures to restrict some views of his activities preclude an officer's observation from a public vantage point where he has a right to be and which renders the activities clearly visible.").

[91] United States v. Knotts, 460 U.S. 276, 285 (1983).

[92] *See* Senate Judiciary Hearing, *supra* note 3, at 94 (2013) (written statement Ryan Calo, Law Professor, University of Washington School of Law) ("The Supreme Court has made it clear through a series of decisions in the nineteen-eighties that there is no search for Fourth Amendment purposes if an airplane or helicopter permits officers to peer into your backyard. I see no reason why these precedents would not readily extend to drones."); Joseph J. Vacek, *Big Brother Will Soon Be Watching—Or Will He? Constitutional, Regulatory, and Operational Issues Surrounding the Use of Unmanned Aerial Vehicles in Law Enforcement*, 85 N.D. L. REV. 673 (2009) ("Constitutionally, it seems that aerial surveillance by any method of any area in open view from any legal altitude does not implicate the Fourth Amendment, as long as the technology used to obtain the surveillance technology is in general public use and does not penetrate into the home"); Brandon Nagy, *Why They Can Watch You: Assessing Constitutionality of Warrantless Unmanned Aerial*

Surveillance by Law Enforcement, 29 BERKELEY TECH. L. J. 135 (2014) ("Those concerned that increasing law enforcement UAS operations erode personal privacy at the expense of the Fourth Amendment do, indeed, have much to worry about. Current Supreme Court jurisprudence, although not directly addressing the Fourth Amendment implications of UAS surveillance, suggests that law enforcement UAS use, with certain limits, would be found constitutional if challenged under the Fourth Amendment.").

[93] Kyllo v. United States, 533 U.S. 27, 34 (2001).

[94] United States v. Jones, 132 S. Ct. 945, 954 (2012) (Sotomayor, J., concurring), *id.* at 957 (Alito, J., concurring in the judgment).

[95] *Id.* at 956 (Sotomayor, J., concurring) ("I would take these attributes of GPS monitoring into account when considering the existence of a reasonable societal expectation of privacy in the sum of one's public movements. I would ask whether people reasonably expect that their movements will be recorded and aggregated in a manner that enables the Government to ascertain, more or less at will, their political and religious beliefs, sexual habits, and so on.").

[96] *Id.* at 963-64 (Alito, J., concurring) ("In the pre-computer age, the greatest protections of privacy were neither constitutional nor statutory, but practical. Traditional surveillance for any extended period of time was difficult and costly and therefore rarely undertaken. The surveillance at issue in this case—constant monitoring of the location of a vehicle for four weeks—would have required a large team of agents, multiple vehicles, and perhaps aerial assistance. Only an investigation of unusual importance could have justified such an expenditure of law enforcement resources. Devices like the one used in the present case, however, make long-term monitoring relatively easy and cheap.").

[97] *See* Smith v. Maryland, 442 U.S. 735 (1979).

[98] *Compare* Klayman v. Obama, 957 F. Supp. 2d 1, 35-36 (D.D.C. 2013), *with* ACLU v. Clapper, 959 F. Supp. 2d 724 (S.D.N.Y. 2013).

[99] *See* Orin S. Kerr, *The Fourth Amendment and New Technologies: Constitutional Myths and the Case for Caution*, 102 MICH. L. REV. 801, 858 (2004).

[100] *See* City of Ontario v. Quon, 560 U.S. 746, 760 (2010) (assuming but not holding that individuals have a reasonable expectation privacy in electronic communications).

[101] *Id.* at 759 ("The Court must proceed with care when considering the whole concept of privacy expectations in communications made on electronic equipment owned by a government employer. The judiciary risks error by elaborating too fully on the Fourth Amendment implications of emerging technology before its role in society has become clear.").

[102] *Jones*, 132 S. Ct. at 945 (Alito, J., concurring in the judgment) ("In circumstances involving dramatic technological change, the best solution to privacy concerns may be legislative. A legislative body is well situated to gauge changing public attitudes, to draw detailed lines, and to balance privacy and public safety in a comprehensive way.") (internal citation omitted).

[103] *Jones*, 132 S.Ct. at 949.

[104] Florida v. Jardines, 133 S. Ct. 1409, 1417-18 (2013).

[105] *See* John Villasenor, *Observations from Above: Unmanned Aircraft Systems and Privacy*, 36 HARV. J. L. & PUB. POL'Y 457, 501 (2013) ("Although privacy expectations are greatly reduced outside the home, the non-governmental use of a UAS to capture images and other information taken while the individual is in a public setting could nonetheless constitute an invasion of privacy.").

[106] *See* Senate Hearing, *supra* note 3, at 71 (statement of Ryan Calo, Professor, University of Washington School of Law) ("[T]here are even fewer limitations on the use of drones by

individuals, corporations, or the press. The common law privacy torts such as intrusion upon seclusion tend to track the constitutional doctrine that there should be no expectation of privacy in public.").

[107] *See* Louis D. Brandeis & Samuel D. Warren, *The Right to Privacy*, 4 HARV. L. REV. 193, 205 (1890).

[108] RESTATEMENT (SECOND) OR TORTS §§652B-652E.

[109] *Id.* §652B.

[110] *See* Shorter v. Retail Credit Co., 251 F. Supp. 329, 322 (D.S.C. 1966).

[111] RESTATEMENT (SECOND) OF TORTS §652B (emphasis added).

[112] N.O.C., Inc. v. Schaefer, 484 A.2d 729, 733 (N.J. Super. Ct. Law Div. 1984).

[113] RESTATEMENT (SECOND) OF TORTS §652B cmt. d.

[114] *See, e.g.*, Miller v. National Broadcasting Co., 187 Cal. App. 3d 1463 (Cal. Ct. App. 1986) (videotaping man in his home while being resuscitated after having suffered a heart seizure); Nader v. General Motors Corp., 25 N.Y.2d 560, 570 (1970) (surveilling plaintiff in bank in an "overzealous" manner).

[115] RESTATEMENT (SECOND) OF TORTS §652B.

[116] *See* 2 DAN B DOBBS ET AL., THE LAW OF TORTS §29 (2d ed. 2011).

[117] *See, e.g.,* DeAngelo v. Fortney, 515 A.2d 594, 596 (Pa. Sup. 1986); Burns v. Masterbrand Cabinets, Inc., 369 Ill. App. 3d 1006, 1012 (Ill. App. Ct. 2007).

[118] RESTATEMENT (SECOND) OF TORTS §652D.

[119] *See id.* at cmt. b.

[120] *Id.*

[121] *Id.* at cmt. a, d.

[122] *See, e.g.*, Wolfson v. Lewis, 924 F. Supp. 1413 (E.D. Pa. 1996) (recording plaintiff's home with sophisticated sound and video equipment); Souder v. Pendleton Detectives, 88 So.2d 716, 718 (La. Ct. App. 1956) (peeping into plaintiff's windows); Egan v. Schmock, 93 F. Supp. 2d 1090, 1094-95 (N.D. Cal. 2000) (filming plaintiff and family while in their home); *see also* RESTATEMENT (SECOND) OF TORTS §652B cmt. b, illus. 2.

[123] *See* Schifano v. Green County, 624 So. 2d 178 (Ala. 1993) (dismissing claim against race track operators who took picture of plaintiffs in winner's circle); Fogel v. Forbes, 500 F. Supp. 1081, 1084, 1087 (E.D. Pa. 1980) (dismissing claim for photographs taken at airport); Tellado v. Time-Life, 643 F. Supp. 904, 907 (D.N.J. 1986) (dismissing claim for photographs taken on battlefield in Vietnam); International Union v. Garner, 601 F. Supp. 187, 191-92 (M.D. Tenn. 1985) (dismissing claim for photographing license plates); Tedeschi v. Reardon, 5 F. Supp. 2d 40, 46 (D. Mass. 1998) (same); Jackson v. Playboy Enterprises, Inc., 574 F. Supp. 10, 13 (S.D. Ohio 1983) (dismissing claim for photographing person on sidewalk); *see also* RESTATEMENT (SECOND) OF TORTS §652B cmt. c (commenting that there is generally no liability for photographing or observing a person while in public "since he is not then in seclusion, and his appearance is public and open to the public eye."); William M. Prosser, *Privacy*, 48 CAL. L. REV. 383, 392 (1960) ("On the public street, or in any other public place, the plaintiff has no right to be alone, and it is no invasion of his privacy to do no more than follow him about. Neither is it such an invasion to take a photograph in such a place, since this amounts to nothing more than making a record, not differing essentially from a full written description, of a public sight which anyone present would be free to see.").

There have been some successful claims for intrusion upon seclusion involving surveillance conducted in public, *see* Kramer v. Downey, 684 S.W. 2d 524, 525 (Tex. Ct. App. 1984) ("[W]e now hold that the right to privacy is broad enough to include the right to be free of those willful intrusions into one's personal life at home and at work which occurred in this case."); Daily Times Democrat v. Graham, 276 Ala. 380, 381 (1964); Huskey v. National Broadcasting Co., Inc., 632 F. Supp. 1282, 1285 (N.D. Ill. 1986); *see also* RESTATEMENT (SECOND) OF TORTS §652B cmt. c ("Even in a public place, however, there may be some matters about the plaintiff, such as his underwear or lack of it, that are not exhibited to the public gaze, and there may still be invasion of privacy when there is intrusion upon these matters."), although this is limited to highly embarrassing situations and appears to be the exception rather than the norm. *See* Jennifer R. Scharf, *Shooting for the Stars: A Call for Federal Legislation to Protect Celebrities' Privacy Rights*, 3 BUFF. INTELL. PROP. L.J. 164, 183 (2006) ("Modifying intrusion to apply in public places would be necessary in order to provide any relief.").

124 *See* McClain v. Boise Cascade Corp., 271 Or. 549, 556 (1975) (dismissing claim where insurance investigator's entered property to take photographs of plaintiff); Mojica Escobar v. Roca, 926 F. Supp. 30, 32-33 (D.P.R. 1996) (dismissing claim against newspaper for photographing outside of home); GTE Mobilnet of South Texas, LTD. Partnership v. Pascouet, 61 S.W. 3d 599, 605 (Tex. Ct. App. 2001) (dismissing claim against company for cell tower workers looking onto property when they serviced nearby cell tower); Aisenson v. American Broadcasting Co, 220 Cal. App. 3d 146, 162-63 (1990) (holding that broadcast of plaintiff while in his driveway and car was not an intrusion upon seclusion); Wehling v. Columbia Broadcasting System, 721 F.2d 506, 509 (5th Cir. 1983) (holding that broadcast of the outside of plaintiff's home taken from public street was not an invasion of privacy); Munson v. Milwaukee Bd. of School Directors, 969 F.2d 266, 271 (7th Cir. 1992) (same).

125 Boring v. Google, Inc., No. 09-2350, 362 Fed. Appx. 273 (3d Cir. 2013).

126 *Id.* at 279.

127 *Id.*

128 Senate Judiciary Hearing, *supra* note 3, at 51 (answer from Michael Toscano, President and Chief Executive Officer, Association of Unmanned Vehicle Systems International).

129 P.L. 112-95, §332, 126 Stat. 123-24.

130 *See, e.g.*, Press Release, Association for Unmanned Vehicle Systems International, AUVSI to FAA: Focus on your Mission, Proceed with UAS Integration (Nov. 28, 2012) ("As an industry, we support a continued, civil dialogue on privacy, but any such conversations should take place concurrent with the integration. The selection process for the six test sites are a separate issue and should be treated as such. Meanwhile, the FAA should adhere to its mission and do what it does best – focus on the safety of the U.S. airspace – while other, more appropriate institutions consider privacy issues."), *available at* http://www.auvsi.org/AUVSINews/AssociationNews.

131 *See* Keith Laing, *Markey: FAA Drone Plan 'Falls Far Short*,' THE HILL (Nov. 7, 2013), *available at* http://thehill.com/policy/transportation/189603-markey-faa-drone-plan-falls-far-short; Electronic Frontier Foundation, The FAA Creates Thin Privacy Guidelines for the Nation's First Domestic Drone "Test Sites," (Dec. 10, 2013), *available at* https://www.eff.org/deeplinks/2013/12/faa-creates-thin-privacy-guidelines-nations-first-domestic-drone-test-sites.

132 *See* FEDERAL AVIATION ADMINISTRATION, INTEGRATION OF CIVIL UNMANNED AIRCRAFT SYSTEMS (UAS) IN THE NATIONAL AIRSPACE SYSTEM (NAS) ROADMAP 11 (Nov. 7, 2013), *available at* https://www.faa.gov/uas/legislative_programs/uas_roadmap/media/UAS_Roadmap_2013.pdf.

133 P.L. 112-95, §332(c), 126 Stat. 73-4 (2012).

134 Unmanned Aircraft System Test Site Program, 78 Fed. Reg. 68360 (Nov. 14, 2013).

[135] *Id.* at 68,364.

[136] *Id.* at 68,361 ("The FAA's authority for including the Final Privacy Requirements in the Test Site OTAs is set forth in 49 U.S.C. §106(l)(6). That statute authorizes the FAA Administrator to enter into an OTA 'on such terms and conditions as the Administrator may consider appropriate.' The FAA believes that it is appropriate to comply with the Final Privacy Requirements.").

[137] *Id.* at 68,632 ("[T]he FAA believes that Test Site operators will be responsive to local stakeholders' privacy concerns and will develop privacy policies appropriately for each Test Site. ... The FAA expects that [the test site operators] will be responsive to stakeholders concerns.").

[138] Notice of Proposed Rulemaking, Operation and Certification of Small Unmanned Aircraft Systems, Federal Aviation Administration (Feb. 15, 2015), *available at* http://www.faa.gov/regulations_policies/rulemaking/recently_published/media/2120-AJ60_NPRM_2-15-2015_joint_signature.pdf.

[139] *Id.* at 36.

[140] Federal Aviation Administration, Integration of Civil Unmanned Aircraft Systems (UAS) in the National Airspace System (NAS) Roadmap (Nov. 7, 2013), *available at* https://www.faa.gov/uas/legislative_programs/uas_roadmap/media/UAS_Roadmap_2013.pdf.

[141] Federal Aviation Administration, Unmanned Aircraft Systems (UAS) Comprehensive Plan: A Report on the Nation's UAS Path Forward 7 (Sept. 2013), *available at* https://www.faa.gov/about/office_org/headquarters_offices/ agi/reports/media/UAS_Comprehensive_Plan.pdf.

[142] Presidential Memorandum, *supra* note 2.

[143] Beyond privacy, the President's memorandum requires that agencies shall ensure they have policies in place that prohibit the collection, use, retention, and dissemination of data in a manner that would violate the First Amendment or would discriminate against persons based upon their ethnicity, race, gender, national origin, religion, sexual orientation, or gender identity, in violation of law. *See* Presidential Memorandum, *supra* note 2.

[144] Presidential Memorandum, *supra* note 8.

[145] Privacy Impact Assessment, *supra* note 29.

[146] 44 U.S.C. §3501 (note).

[147] 6 U.S.C. §142.

[148] 6 U.S.C. §142(a)(5).

[149] Dep't of Homeland Security, Working Group to Safeguard Privacy, Civil Rights, and Civil Liberties in the Department's Use and Support of Unmanned Aerial Systems (Sept. 14, 2012), *available at* https://www.dhs.gov/sites/ default/files/publications/foia/working-group-to-safeguard-privacy-civil-rights-and-civil-liberties-in-the-departmentsuse-and-support-of-unmanned-aerial-systems-uas-s1-information-memorandum-09142012.pdf.

[150] DEP'T OF HOMELAND SECURITY, PRIVACY OFFICE, 2014 ANNUAL REPORT TO CONGRESS 19 (Sept. 30, 2014), *available at* http://www.dhs.gov/sites/default/files/publications/dhs-privacy-office-2014-annual-report-FINAL.pdf.

[151] DOJ Inspector General Report, *supra* note 33, at ii.

[152] *Id.*

[153] Harley Geiger, White House Drone Memo Right to Focus on Privacy, Center for Democracy and Technology (Feb. 15, 2015), *available at* https://cdt.org/press/white-house-drone-memo-right-to-focus-on-privacy/.

[154] *Id.*

[155] S. 1639, 113th Cong. (2013); H.R. 2868, 113th Cong. (2013).

[156] H.R. 972, 113th Cong. (2013). Senator Rand Paul filed similar legislation in the 112th Congress, but did not reintroduce it in the 113th Congress. *See* S. 3287, 112th Cong. (2012). Unlike H.R. 972, S. 3287 included an express exclusionary rule for evidence obtained in violation of the act.

[157] H.R. 637, 113th Cong. (2013).

[158] H.R. 637, 113th Cong. (2013).

[159] Letter from Senator Rand Paul to Federal Bureau of Investigation Director Robert Mueller (July 25, 2013), *available at* http://www.paul.senate.gov/files/documents/072513 Paulresponse.pdf.

[160] Letter from Stephen D. Kelly, Assistant Director, Office of Congressional Affairs, Dep't of Justice to Senator Rand Paul (July 29, 2013), *available at* http://www.paul.senate.gov/files/documents/072913FBIResponse.pdf.

[161] New State Ice Co. v. Liebmann, 285 U.S. 262 (1932) (Brandeis, J., dissenting).

[162] *See* Bennett, *supra* note 12; Kaminski, *supra* note 12.

[163] National Conference of State Legislatures, 2014 Unmanned Aircraft Systems (UAS) Legislation, *available at* http://www.ncsl.org/research/civil-and-criminal-justice/2014-state-unmanned-aircraft-systems-uas-legislation.aspx

[164] *See, e.g.*, ALASKA STAT. §18.65.902; FLA. STAT. ANN. §934.50.

[165] 725 ILL. COMP. STAT. §167/20.

[166] ALASKA. CODE §18.65.902.

[167] IDAHO CODE §21-213.

In: Drones in the U.S.
Editor: Sophia Barrett

ISBN: 978-1-63484-918-0

Chapter 2

TESTIMONY OF AMIE STEPANOVICH, DIRECTOR OF THE DOMESTIC SURVEILLANCE PROJECT, ELECTRONIC PRIVACY INFORMATION CENTER. HEARING ON "THE FUTURE OF DRONES IN AMERICA: LAW ENFORCEMENT AND PRIVACY CONSIDERATIONS"*

Mister Chairman and Members of the Committee, thank you for the opportunity to testify today concerning the use of drones by law enforcement in the United States. My name is Amie Stepanovich. I am the Director of the Domestic Surveillance Project at the Electronic Privacy Information Center.

EPIC is a non-partisan research organization, established in 1994, to focus public attention on emerging privacy and civil liberties issues.[1] We work with a distinguished panel of advisors in the fields of law, technology, and public policy.[2] We have a particular interest in the protection of individual privacy rights against government surveillance. In the last several years, EPIC has taken a particular interest in the unique privacy problems associated with aerial drones.

The Federal Aviation Administration ("FAA") has been directed to fully integrate drones into the National Airspace by 2015.[3] In 2012 EPIC petitioned

* This is an edited, reformatted and augmented version of testimony presented March 20, 2013 before the Senate Judiciary Committee.

the FAA, as it considers new regulations to permit the widespread deployment of drones, to also develop new privacy safeguards.[4] The FAA heeded our warning, and is now considering privacy policies for drone operators. However, more must be done to protect the privacy of individuals in the United States.

We appreciate the Committee's interest in domestic drone use and its substantial impact on the privacy of individuals in the United States. In my statement today, I will describe the unique threats to privacy posed by drone surveillance, the problems with current legal safeguards, and the need for Congress to act.

I. Aerial Drones Pose a Unique Threat to Privacy

A drone is an aerial vehicle designed to fly without a human pilot on board. Drones can either be remotely controlled or autonomous. Drones can be weaponized and deployed for military purposes.[5] Drones can also be equipped with sophisticated surveillance technology that makes it possible to spy on individuals on the ground. In a report on drones published by EPIC in 2005, we observed, "the use of [drones] gives the federal government a new capability to monitor citizens clandestinely, while the effectiveness of the...surveillance planes in border patrol operations has not been proved."[6] Today, drones greatly increase the capacity for law enforcement to collect personal information on individuals.

We recognize that there are many positive applications for drones within the United States. With little to no risk to individual privacy, drones may be used to combat forest fires, conduct search and rescue operations, survey emergency situations, and monitor hurricanes and other weather phenomena.[7] In Dallas, a drone used by a hobbyist photographer was able to pinpoint an instance of gross environmental abuse at a nearby factory.[8] In Alabama, drones were recently used to assist in monitoring a hostage situation involving a young boy abducted off of the school bus.[9]

However, when drones are used to obtain evidence in a criminal proceeding, intrude upon a reasonable expectation of privacy, or gather personal data about identifiable individuals, rules are necessary to ensure that fundamental standards for fairness, privacy, and accountability are preserved.

The technology in use today is far more sophisticated than most people understand. Cameras used to outfit drones are among the highest definition cameras available. The Argus camera, featured on the PBS Nova documentary on drones, has a resolution of 1.8 gigapixels and is capable of observing objects as small as six inches in detail from a height of 17,000 feet.[10] On some drones, sensors can track up to 65 different targets across a distance of 65 square miles.[11] Drones may also carry infrared cameras, heat sensors, GPS, sensors that detect movement, and automated license plate readers.[12]

Recent records received by EPIC under the Freedom of Information Act demonstrate that the Bureau of Customs and Border Protection procured drones outfitted with technology for electronic signals interception and human identification.[13] Following receipt of these documents, EPIC and a broad coalition of privacy and civil liberties organizations petitioned the CBP to suspend the domestic drone program, pending the establishment of privacy safeguards.[14]

Much of this surveillance technology could, in theory, be deployed on manned vehicles. However, drones present a unique threat to privacy. Drones are designed to maintain a constant, persistent eye on the public to a degree that former methods of surveillance were unable to achieve. Drones are cheaper to buy, maintain, and operate than helicopters, or other forms of aerial surveillance.[15] Drone manufacturers have recently announced new designs that would allow drones to operate for more than 48 consecutive hours,[16] and other technology could extend the flight time of future drones into spans of weeks and months.[17] Also, "by virtue of their design, size, and how high they can fly, [drones] can operate undetected in urban and rural environments."[18]

Drones are currently being developed that will carry facial recognition technology, able to remotely identify individuals in parks, schools, and at political gatherings.[19] The ability to link facial recognition capabilities on drones operated by the Department of Homeland Security ("DHS") to the Federal Bureau of Investigation's Next Generation Identification database or DHS' IDENT database, two of the largest collections of biometric data in the world, further exacerbates the privacy risks.[20]

Law enforcement offices across the country have expressed interest in the purchase and use of drone technology to assist with law enforcement operations. Records released in 2012 by the Federal Aviation Administration show that over 220 public entities have already received approval to operate drones over the United States, including Police departments from Texas, Kansas, Washington, and other states.[21] The Florida Police Chiefs Association expressed a desire to use drones to conduct general crowd surveillance at public events.[22] News reports demonstrate that other police departments are not only interested in invasive surveillance equipment, but have also voiced interest in outfitting drones with non-lethal weapons.[23]

II. Current Privacy Safeguards Are Inadequate

The Supreme Court has not yet considered the limits of drone surveillance under the Fourth Amendment, though the Court held twenty years ago that law enforcement may conduct manned aerial surveillance operations from as low as 400 feet without a warrant.[24] In addition, no federal statute currently provides adequate safeguards to protect privacy against increased drone use in the United States. Accordingly, there are substantial legal and constitutional issues involved in the deployment of aerial drones by law enforcement and state and federal agencies that need to be addressed. Technologist and security expert Bruce Schneier observed earlier this year at an event hosted by EPIC on Drones and Domestic Surveillance, "today's expensive and rare is tomorrow's commonplace."[25] As drone technology becomes cheaper and more common, the threat to privacy will become more substantial. High-rise buildings, security fences, or even the walls of a building are not barriers to increasingly common drone technology.

The Supreme Court is aware of the growing risks to privacy resulting from new surveillance technology but has yet to address the specific problems associated with drone surveillance. In *United States v. Jones*, a case that addressed whether the police could use a GPS device to track the movement of a criminal suspect without a warrant, the Court found that the installation and deployment of the device was an unlawful search and seizure.[26] Justice Sotomayor in a concurrence pointed to broader problems associated with new forms of persistent surveillance.[27] And Justice Alito, in a separate concurrence joined by three other Justices, wrote, "in circumstances involving dramatic technological change, the best solution to privacy concerns may be legislative."[28]

Regarding the invasive use of drones by commercial operators, current law does not anticipate the use of mobile devices that can hover outside a bedroom window or follow a person down a street. Legal standards should be established to protect people from a violation of reasonable expectations of privacy, including surveillance in public spaces. In consideration legislation to address law enforcement use of drones, it would be appropriate also to establish privacy standards for the commercial use of drones.

III. CONGRESS SHOULD ESTABLISH SAFEGUARDS RELATED TO THE USE OF DRONES

As the Chairman has indicated, the privacy and security concerns arising from the use of drones needs to be addressed.[29] In order to mitigate the risk of increased use of drones in our domestic skies, Congress must pass targeted legislation, based on principles of transparency and accountability.

State and local governments have considered a wide array of laws and regulations to prevent abuses associated with drone technology.[30] A current survey demonstrates that over 30 states have proposed legislation to protect against unregulated drone surveillance of individuals.[31] Most of these bills mandate a warrant requirement for the collection of information by drones operated by law enforcement officials.[32] Other bills require reporting requirements for drone operators.[33] A bill in Georgia restricts law enforcement use of drones strictly to felony investigations,[34] and a bill circulating in Oregon would require state approval for all drones, including federal drones, that would fly over the state's airspace.[35]

Even as states consider these various measures, it would be appropriate for Congress to establish privacy standards for the operation of drones in the United States. First, Congress should require all drone operators, both public and commercial, to submit, prior to receipt of a drone license, a detailed report on the drones' intended use. This report should describe, the specific geographic are where the drone will be deployed, the mission that the drone is expected to fulfill, and the surveillance equipment with which the drone will be outfitted. Each of these reports should be made publicly available at a publicly accessible web site. A private right of action and, in certain instances, federal prosecution authority should be included to ensure that drone operators comply with the terms of these statements.

In order to prevent abuses associated with the use of this technology, a strict warrant requirement needs to be implemented for all drone surveillance conduct by law enforcement. A warrant requirement would establish a presumption that evidence obtained by means of an aerial search should require judicial approval. Statutory exceptions could be created for exigency in order to address drone use in emergency situations or when necessary to protect human life. In addition, mandatory public reporting requirements, similar to those required by the Wiretap Act, would increase the transparency and accountability of law enforcement drone operations.[36]

Ongoing surveillance of individuals by aerial drones operating in domestic airspace should be prohibited. The invasiveness of drone technology represents a privacy risks to individuals as they pursue their daily activities. A drone, with the capability of staying aloft for hours or days at a time, could monitor a person's entire life as they go from home to work to school to the store and back. Even if law enforcement is not able to immediately discern exactly what a person says or does or buys at a particular location, simply tracking an individual's public movements in a systematic fashion for extended periods of time can create a vivid description of their private life.[37] Broad, unregulated drone surveillance would have a chilling effect on the speech and expression rights of individuals in the United States. Drones should not be used as robotic patrol officers for law enforcement.

Finally, drone surveillance technology may allow the collection of information and images that would otherwise be inaccessible to prying eyes, such as activities within the home. Congress should prohibit drone operators from conducting surveillance of individuals that infringes on property rights. A federal "Peeping Tom" statute, recognizing the enhanced capabilities of aerial drones, would provide baseline privacy protection for individuals within the home. Additional provisions should prevent against any use of drones to collect information that would not otherwise be retrievable without a physical trespass.

Additional drone legislation should include:

- Use Limitations – Prohibitions on general surveillance that limit law enforcement drone surveillance to specific, enumerated circumstances, such as in the case of criminal surveillance subject to a warrant, a geographically---confined emergency, or for reasonable non---law enforcement use where privacy will not be substantially affected;
- Data Retention Limitations – Restrictions on retaining or sharing surveillance data collected by drones, with emphasis on personally identifiable information;
- Transparency and Public Accountability – A requirement for all federal agencies that choose to operate drones to promulgate privacy regulations, subject to the notice and comment provisions of the Administrative Procedure Act. In addition, the law should provide for third party audits and oversight for law enforcement drone operations.

These three principles would further help protect the privacy interests of individuals against both government and commercial drone operators.

CONCLUSION

The increased use of drones to conduct surveillance in the United States must be accompanied by increased privacy protections. The current state of the law is insufficient to address the drone surveillance threat. EPIC supports legislation aimed at strengthening safeguards related to the use of drones as surveillance tools and allowing for redress for drone operators who fail to comply with the mandated standards of protection. We also support compliance with the Administrative Procedure Act for the deployment of drone technology and limitations for federal agencies and other organizations that initially obtain a drone for one purpose and then wish to expand that purpose.

Thank you for the opportunity to testify today. I will be pleased to answer your questions.

End Notes

[1] About EPIC, EPIC, http://www.epic.org/about (last visited July 16, 2012).

[2] EPIC Advisory Board, EPIC, http://www.epic.org/epic/advisory_board.html (last visited July 16, 2012).

[3] Federal Aviation Administration Modernization and Reform Act of 2012 ("FMRA"), Pub. L. 112-95 §§ 331- 336 (2012), available at http://www.gpo.gov/fdsys/pkg/PLAW-112publ95/pdf/PLAW-112publ95.pdf.

[4] Unmanned Aerial Vehicles (UAVs) and Drones, EPIC, http://www.epic.org/privacy/drones (last visited July 16, 2012).

[5] See, e.g., Predator B UAS, General Atomics Aeronautical, http://www. ga-asi.com/products/aircraft/predator_b.php (last visited June 25, 2012); X--47B UCAS, Northrop Grumman, http://www.as.northropgrumman.com/products/nucasx47b/index.html (last visited July 16, 2012).

[6] Spotlight on Surveillance: Unmanned Planes Offer New Opportunities for Clandestine Government Tracking (August 2005), EPIC, http://epic.org/privacy/surveillance/spotlight/0805/ (last visited July 16, 2012).

[7] See, e.g., Tim Wall, Flying Drones Fight Fires, Discovery News (Nov. 10, 2011), available at http://news.discovery.com/earth/flying-drones-fight-fires-111110.html; Meghan Keneally, Drone Plane Spots a River of Blood Flowing From the Back of a Dallas Meat Packing Plant, Daily Mail Online (Jan. 24, 2012), available at http://www.dailymail.co.uk/news/article-2091159/A-drone-plane-spots-river-blood-flowingDallas-meat-packing-plant.html; Sean Holstege, Drones' Good Flies Hand in Hand with Bad, Experts Fear,

AZCentral (July 7, 2012), available at http://www.azcentral.com/12news/news/articles/2012/07/07/20120707arizona-unmanned-dronesconcerns.html.

[8] Meghan Keneally, Drone Plane Spots a River of Blood Flowing From the Back of a Dallas Meat Packing Plant, Daily Mail Online (Jan. 24, 2012), available at http://www.dailymail.co.uk/news/article-2091159/A-droneplane-spots-river-blood-flowing-Dallas-meat-packing-plant.html.

[9] See Military Tactics, Equipment Helped Authorities End Alabama Hostage Standoff, Fox News (Feb. 7, 2013), http://www.foxnews.com/us/2013/02/07/alabama-kidnapper-was-killed-in-firefight-during-stormingbunker-fbi-says/.

[10] Ryan Gallagher, Could the Pentagon's 1.8 Gigapixel Drone Camera Be Used for Domestic Surveillance, Slate (Feb. 6, 2013), http://www.slate.com/blogs/future_tense/2013/02/06/argus_is_could_the_pentagon_s_1_8_gigapixel_drone _camera_be_used_for_domestic.html.

[11] Id.

[12] Customs and Border Protection Today, Unmanned Aerial Vehicles Support Border Security (July/Aug. 2004), available at http://www.cbp.gov/xp/CustomsToday/2004/ Aug/other/aerial_vehicles.xml.

[13] Declan McCullagh, DHS Built Domestic Surveillance Tech into Predator Drones, CNET (Mar. 2, 2013), http://news.cnet.com/8301---13578_3---57572207---38/dhs---built---domestic---surveillance---tech---into---predator--- drones/.

[14] Letter from the Electronic Privacy Information Center, et al. to David V. Aguilar, Deputy Commissioner, U.S. Bureau of Customs and Border Protection (Mar. 19, 2013), available at http://epic.org/drones_petition/.

[15] Nick Wingfield and Somini Sengupta, Drones Set Sights on U.S. Skies, NY Times (Feb. 17, 2012), available at http://www.nytimes.com/2012/02/18/technology/drones---with---an---eye---on---the---public---cleared---to---fly.html?pagewanted=all; http://www.wired.com/autopia/2012/05/drone---auto---vids/; Sabrina Hall, Shelby County Sheriff's Department Wants Drones, WREG (May 3, 2012), available at http://wreg.com/2012/05/03/shelby---county---sheriffs---department---wants---drones/. Drones can run from $300 for the most basic drone, able to record and transmit video, to $18 million for a General Atomics Predator B drone, the model owned by the United States Bureau of Customs and Border Protection. See Parrot AR.Drone 2.0, Apple, http://store.apple.com/us/product/H8859ZM/A (last visited July 16, 2012); Office of the Inspector Gen., Dep't Homeland Security, OIG---12---85, CBPs Use of Unmanned Aircraft Systems in the Nation's Border Security (May 2012), available at http://www.oig.dhs.gov/assets/Mgmt/2012/OIG_12---85_May12.pdf [hereinafter DHS OIG Report] at 2.

[16] Mark Brown, Lockheed Uses Ground--Based Laser to Recharge Drone Mid--Flight (July 12, 2012), available at http://www.wired.co.uk/news/archive/2012---07/12/lockheed---lasers.

[17] Steven Aftergood, Secret Drone Technology Barred by "Political Conditions" (Mar. 22, 2012), available at http://www.fas.org/blog/secrecy/2012/03/sandia_drone.html.

[18] Jennifer Lynch, Are Drones Watching You?, Electronic Frontier Foundation (Jan. 10, 2012), available at https://www.eff.org/deeplinks/2012/01/drones---are---watching---you.

[19] Clay Dillow, Army Developing Drones that Can Recognize Your Face From a Distance, PopSci (Sept. 28, 2011, 4:01 PM), http://www.popsci.com/technology/article/2011---09/army---wants---drones---can---recognize---your--- face---and---read---your---mind.

[20] See Next Generation Identification, Federal Bureau of Investigation, http://www.fbi.gov/about--- us/cjis/fingerprints_biometrics/ngi/ngi2/ (last visited July 16, 2012); Privacy Impact Assessment, Department of Homeland Security, Automated Biometric Identification

System (IDENT) (July 31, 2006), http://www.dhs.gov/xlibrary/assets/privacy/privacy_pia_usvisit_ident_final.pdf.

[21] See letter from Michael Huerta, Acting Administrator, Federal Aviation Administration to the Honorable Edward J. Markey (Sept. 21, 2012), available at http://markey.house.gov/sites/markey.house.gov/files/documents/FAA%20drones%20response.pdf; see also Jennifer Lynch, Just How Many Drone Licenses Has the FAA Really Issued, Electronic Frontier Foundation (Feb. 21, 2013), https://www.eff.org/deeplinks/2013/02/just-how-many-drone-licenses-has-faa-reallyissued (providing details on contradictory statements made by the Federal Aviation Adminstration regarding the issuance of drone licenses).

[22] See Florida Ban on Drones Advances Despite Law Enforcement Objections, Fox News (Feb. 7, 2013), http://www.foxnews.com/politics/2013/02/07/fla-police-want-to-use-drones-for-crowd-control/.

[23] See Conor Friedersdorf, Congress Should Ban Armed Drones Before Cops in Texas Deploy One, the Atlantic (May 24, 2012), http://www.theatlantic.com/national/archive/2012/05/congress-should-ban-armeddrones-before-cops-in-texas-deploy-one/257616/.

[24] See Florida v. Riley, 488 U.S. 445 (1989) (holding that a police helicopter flying more than 400 feet above private property is not a search).

[25] Drones and Domestic Surveillance, EPIC, http://epic.org/events/drones/ (last visited Mar. 15, 2013).

[26] United States v. Jones, 132 S.Ct. 945, 949 (2012). See also U.S. v. Jones, EPIC, http://epic.org/amicus/jones/.

[27] Id. at 954-57.

[28] Id. at 964.

[29] Press Release from Senator Patrick Leahy, The Agenda of the Senate Judiciary Committee for the 113th Congress (Jan. 16, 2013), available at http://www.leahy.senate.gov/press/113-sjc-agenda-speech (Iam concerned about the growing use of drones by federal and local authorities to spy on Americans here at home. This fast-emerging technology is cheap and could pose a significant threat to the privacy and civil liberties of millions of Americans.”).

[30] See, e.g., Erika Neddenien, ACLU Teams with Lawmaker to Push Regulation of Unmanned Drones in VA, WTVR (July 12, 2012http://wtvr.com/2012/07/12/aclu-working-with-lawmaker-to-push-regulation-of-unmanneddrones-in-va/ (last visited July 16, 2012); Press Release, Seattle City Council, Seattle City Council Committee to Discuss Drones in Seattle and the Issues they Present (May 1, 2012), available at http://council.seattle.gov/2012/05/01/seattle-city-council-committee-to-discuss-drones-in-seattle-and-theissues-they-present/.

[31] Allie Bohm, Status of Domestic Drone Legislation in the States, American Civil Liberties Union (Mar. 14, 2013), http://www.aclu.org/blog/technology-and-liberty/status-domestic-drone-legislation-states.

[32] Allie Bohm, Drone Legislation:What’s Being Proposed in the States?, American Civil Liberties Union (Mar. 6, 2013), http://www.aclu.org/blog/technology-and-liberty-national-security/drone-legislation-whats-beingproposed-states (Noting that states that have introduced a bill to require a warrant for police drone surveillance include Arizona, California, Florida, Georgia, Idaho, Illinois, Kentucky, Maryland, Massachusetts, Minnesota, Missouri, Montana, New Hampshire, New Mexico, North Dakota, Oklahoma, Oregon, Rhode Island, South Carolina, Tennessee, Texas, Washington, and Wyoming.).

[33] See id. (Noting that states that have introduced a bill that includes a reporting requirement include Hawaii, Illinois, Maine, Mass, Rhode Island, Washington.).

[34] See id.

[35] Oregon SB 71 (2013), available at www.leg.state.or.us/13reg/measures/sb0001.dor/sb0071.intro.html.

[36] See 18 U.S.C. § 2519.

[37] See EPIC: Locational Privacy, https://epic.org/privacy/location_privacy/default.html.

In: Drones in the U.S.
Editor: Sophia Barrett

ISBN: 978-1-63484-918-0

Chapter 3

STATEMENT OF RYAN CALO, ASSISTANT PROFESSOR, UNIVERSITY OF WASHINGTON SCHOOL OF LAW. HEARING ON "THE FUTURE OF DRONES IN AMERICA: LAW ENFORCEMENT AND PRIVACY CONSIDERATIONS"*

Thank you Chairman Leahy, ranking Member Grassley, and Members of the Committee for this opportunity to testify today.

My name is Ryan Calo and I am a law professor at the University of Washington. I am also the former director for privacy and robotics at the Stanford Law School Center for Internet and Society.

Last year, Congress charged the Federal Aviation Administration (FAA) with accelerating the integration of unmanned aircraft systems—known colloquially as "drones"—into domestic airspace.[1] Drones are not new; we deployed them for target practice throughout World War II.[2] What is new is the prospect of their widespread use over American cities and towns.

Drones have a lot of people worried about privacy—and for good reason. Drones drive down the cost of aerial surveillance to worrisome levels. Unlike fixed cameras, drones need not rely on public infrastructure or private partnerships. And they can be equipped not only with video cameras and

---

* This is an edited, reformatted and augmented version of a statement presented March 20, 2013 before the Senate Judiciary Committee.

microphones, but also the capability to sense heat patterns, chemical signatures, or the presence of a concealed firearm.

American privacy law, meanwhile, places few limits on aerial surveillance. We enjoy next to no reasonable expectation of privacy in public, or from a public vantage like the nation's airways. The Supreme Court has made it clear through a series of decisions in the nineteen-eighties that there is no search for Fourth Amendment purposes if an airplane or helicopter permits officers to peer into your backyard.[3] I see no reason why these precedents would not extend readily to drones.

Drones may also follow people around from place to place, even after the recent decision of *United States v. Jones.*[4] *Jones* held that affixing a global positioning device to a vehicle for the purpose of tracking the location of the occupant is a search within the meaning of the Fourth Amendment. But it is far from certain how *Jones* would apply to surveillance by a drone, which need not be affixed to anything.

Citizens have no reasonable expectation of privacy in contraband. Dogs can sniff luggage or cars without triggering the Fourth Amendment because, courts assume, dogs only alert in the presence of narcotics or other illegal possessions.[5] A logical extension of this precedent, it seems to me, is that drones could fly around testing the air for drug particles and report back suspicious activity to law enforcement without ever implicating the Constitution.[6]

I have heard it suggested that the Supreme Court's decision in *Kyllo v. United States* involving thermal imaging limits how drones might be used for surveillance. *Kyllo* holds, in essence, that officers need probable cause to peer into the home using technology that is unavailable to the general public.[7] Setting aside whether drones would even draw a *Kyllo* analysis, the technology will indeed be available to the general public as soon as 2015 when the FAA relaxes its ban on commercial use.

The subject of today's hearing is drones and law enforcement. I pause only to note that, if anything, there are even fewer limits on the use of drones by individuals, corporations, or the press. The common law privacy torts such as intrusion upon seclusion tend to track the constitutional doctrine that there should be no expectation of privacy in public.[8] Some might go further and argue that the press (at least) has a free speech interest in using technology to cover newsworthy events.[9]

This combination of cheap, powerful surveillance and inadequate privacy law has understandably resulted in a backlash against drones, one further compounded by our association of the technology with the theatre of war.

This is in many ways a shame. Drones have the potential to be a transformative technology, helping governments, empowering civilians, and fostering innovation in countless ways. As the Congressional Research Service recently stated in a report, "the extent of [drone's] potential domestic application is bound only by human ingenuity."[10] Drones can be lifesavers in the hands of police and firefighters and flying smart phones in the hands of consumers and private industry.

I am very concerned that we will not realize the potential of this technology because we have been so remiss in addressing the legitimate privacy concerns that attend it. There are several ways the government could change this picture. Ideally, we would take the opportunity to finally drag privacy law into the twenty-first century by reexamining our outmoded doctrines. This is a slow process, but courts do seem to be making strides in recent years.

Several federal bills have proposed placing limits on drones. I think we should be very careful here for a few reasons. First, the problem is broader than unmanned aircraft systems: flight is not a prerequisite for threatening civil liberties. There are robots that climb the side of buildings, for instance, that would not be covered under the draft bills I've read. Second, there is likely some benefit to allowing individual states to adopt different approaches to drones and seeing what works and what does not.

There is one approach that I believe could act as stop-gap, and that is for Congress to instruct the FAA to take privacy into account as part of its mandate to integrate drones into domestic airspace.[11] The agency has been largely silent on the issue of privacy—only recently did members of the privacy community receive a letter from the FAA asking for input in connection with the selection of drone testing centers.

But the FAA could require public and eventually private applicants to furnish the agency with a plan to minimize their impact on privacy as part of the application. The agency could then consider the plan, and even withdraw the license for those who flout it. This might help allay reasonable concerns over drones in the short term while continuing to permit their innovative and lifesaving uses.

Thank you again for the opportunity to speak today. I look forward to your questions.

End Notes

[1] FAA Modernization and Reform Act of 2012, P.L. 112-95, 126 Stat. 11.

[2] LAWRENCE NEWCOME, UNMANNED AVIATION: A BRIEF HISTORY OF UNMANNED AERIAL VEHICLES 48 (2004).

[3] See Florida v. Riley, 488 U.S. 445 (1989); California v. Ciraolo, 476 U.S. 207 (1986); Dow Chem. Co. v. United States, 476 U.S. 227 (1986).

[4] 132 S.Ct. 945 (2012).

[5] See Illinois v. Caballes, 543 U.S. 405 (2005); United States v. Place, 462 U.S. 696, 707 (1983) ("A 'canine sniff' by a well-trained narcotics detection dog, however, does not ... expose noncontraband items that otherwise would remain hidden from public view.").

[6] See Ryan Calo, The Drone as Privacy Catalyst, 64 STAN. L. REV. ONLINE 29, 31 (2011).

[7] Kyllo v. United States, 533 U.S. 27 (2001).

[8] E.g., RESTATEMENT (SECOND) OF TORTS § 652B. But see Daily Democrat v. Graham, 276 Ala. 380, 381 (1964) (plaintiff—whose dress had been blown up by the wind in a public place— allowed to pursue privacy tort against defendant photographer).

[9] Cf. Glik v. Cunniffe, 655 F.3d 78 (1st Cir. 2011) (holding that a citizen has a First Amendment right to videotape police during course of his arrest). Thank you to Margot Kaminiski for this pointer.

In: Drones in the U.S.
Editor: Sophia Barrett
ISBN: 978-1-63484-918-0

Chapter 4

TESTIMONY OF JOHN VILLASENOR, NONRESIDENT SENIOR FELLOW, THE BROOKINGS INSTITUTION. HEARING ON "EYES IN THE SKY: THE DOMESTIC USE OF UNMANNED AERIAL SYSTEMS"*

Good morning Chairman Sensenbrenner, Ranking Member Scott, and Members of the Subcommittee. Thank you very much for the opportunity to testify today on the important topic of privacy and unmanned aircraft systems (UAS).[1]

I am a nonresident senior fellow in Governance Studies and the Center for Technology Innovation at the Brookings Institution. I am also a professor at UCLA, where I hold appointments in the Electrical Engineering Department and the Department of Public Policy. The views I am expressing here are my own, and do not necessarily represent those of the Brookings Institution or the University of California. Portions of my testimony today are adapted from a law review article I recently published in the *Harvard Journal of Law and Public Policy*.[2]

UAS, often referred to as "drones," can be employed in an endless variety of civilian applications, the overwhelming majority of them beneficial. However, like any technology, UAS can also be misused. The most common concern regarding domestic UAS relates to their potential impact on privacy.

---

* This is an edited, reformatted and augmented version of testimony presented May 17, 2013 before the House Judiciary Committee, Subcommittee on Crime, Terrorism, Homeland Security and Investigations.

This is a legitimate concern. Existing laws and jurisprudence provide an important foundation, but they also leave many questions unanswered.

For non-government operators, determining when UAS use violates privacy involves the tension between First Amendment freedoms and common law and statutory privacy protections. With respect to government-operated UAS, the Fourth Amendment is of course central to the privacy question. While the Supreme Court has never explicitly considered warrantless observations using UAS, a careful examination of Supreme Court privacy jurisprudence suggests that the Constitution will provide a much stronger measure of protection against government UAS privacy abuses than is widely appreciated. The Fourth Amendment has served us well since its ratification in 1791, and there is no reason to suspect it will be unable to do so in a world where unmanned aircraft are widely used.

This does not mean that there is no need for additional statutory UAS privacy protections. However, when drafting new laws it is critical to adopt a balanced approach that recognizes the inherent difficulty of predicting the future of any rapidly changing technology. Although unmanned aircraft pose real and increasingly well-recognized privacy concerns, they also offer real and much less widely understood benefits. A dialog conducted with full awareness of this balance will be much more likely to lead to positive policy outcomes.

UNMANNED AIRCRAFT SYSTEMS

While much of the public and legislative interest in UAS is recent, unmanned flight has a long history. In fact, early research in unmanned aviation laid some of the critical groundwork that was later used by pioneers in manned aviation,[3] including the Wright brothers, who achieved sustained manned heavier-than-air flight in 1903. Unmanned flight received significant attention throughout most of the twentieth century, particularly in the military aviation and model aircraft communities.

The dramatic growth over the last decade in worldwide UAS use is due to a confluence of multiple factors. Continued improvements in electronics have made it possible to equip even very small UAS platforms with sophisticated on-board computational systems for tasks such as image processing and GPS-based navigation. Thanks to advances in digital imaging systems and wireless communications, high resolution images and video acquired by an unmanned aircraft can be transmitted in real time to an observer fifty feet—or 5,000 miles— away. Innovations in airframe design and flight control methods are making it possible to design smaller and more agile UAS.

Unmanned aircraft will dominate the future of aviation as thoroughly as manned aircraft have dominated its past. In the U.S. military, which now has thousands of unmanned aircraft systems, that transition is well underway. Civilian UAS use in the United States is also set to grow rapidly following the enactment of the FAA Modernization and Reform Act of 2012,[4] which provides for the integration of UAS into the national airspace by late 2015.[5]

Unmanned aircraft come in an incredible variety of shapes and sizes. Some, such as the Global Hawk used by the U.S. military, are as large and nearly as fast as business jets.[6] Others are small enough to fit in a backpack or even the palm of a hand. The video-capable Nano Hummingbird, developed by California-based AeroVironment, weighs only two-thirds of an ounce.[7] In May 2013, a team of Harvard researchers reported the successful flight of the RoboBee, a robotic insect weighing less than one three-hundredth of an ounce, and powered by electricity delivered through a thin wire attached to an external power source.[8] In the summer of 2010, a solar-powered Qinetiq Zephyr weighing slightly over 110 pounds despite having a wingspan of about 74 feet stayed aloft for over two continuous weeks over Arizona.[9]

The potential applications of UAS are just as varied. UAS can help rescuers identify people in need of assistance following a natural disaster and provide vital overhead imagery to police officers attempting to defuse a hostage standoff. In the commercial world, UAS will be employed for tasks as diverse as surveying, crop spraying, and traffic monitoring. Scientific applications include air quality assessment, wildlife tracking, and measuring the internal dynamics of violent storms. UAS will also generate a number of economic benefits, both by creating jobs in unmanned aircraft design and production and by spurring advances in robotics that will apply well beyond aviation, in fields ranging from manufacturing to surgery.

It is also undeniable that UAS can be used in ways that violate privacy. For privacy from government UAS, the Fourth Amendment provides the key constitutional foundation.

SUPREME COURT FOURTH AMENDMENT JURISPRUDENCE

The Fourth Amendment, which provides the "right of the people to be secure in their persons, houses, papers, and effects, against unreasonable searches and seizures," has been a cornerstone of privacy from government intrusion for over two centuries. While the Supreme Court has never ruled specifically on when UAS use constitutes a Fourth Amendment search, there is a long list of relevant precedents. Among these are two cases from the 1980s that addressed warrantless observations of the curtilage of a home from manned government aircraft.[10]

The first of these cases, *California v. Ciraolo*,[11] started with a September 1982 tip to police in Santa Clara, California regarding backyard marijuana cultivation. After finding the yard surrounded by high fencing obscuring the view from the street, the police obtained a small airplane and flew over the residence at 1,000 feet. The officers on the airplane observed and photographed what they concluded to be marijuana plants growing in the backyard. This evidence was used to obtain a search warrant to seize the plants.

In a May 1986 decision, the Supreme Court noted that warrantless police observations of curtilage are not necessarily unconstitutional: "The Fourth Amendment protection of the home has never been extended to require law enforcement officers to shield their eyes when passing by a home on public thoroughfares."[12] Moreover, the Court noted, because the observations were made from "public navigable airspace . . . in a physically nonintrusive manner," the respondent's expectation of privacy from such aerial observations was not one "that society is prepared to honor."[13] The Court concluded that "[i]n an age where private and commercial flight in the public airways is routine, it is unreasonable for respondent to expect that his marijuana plants were constitutionally protected from being observed with the naked eye from an altitude of 1,000 feet."[14]

Just under three years after *Ciraolo*, the Supreme Court once again considered the constitutionality of aerial observations of a home's curtilage by law enforcement. Like *Ciraolo*, *Florida v. Riley*[15] arose from a tip involving marijuana cultivation behind a house. In *Riley*, the observations were made from a helicopter at 400 feet, enabling officers to see the plants through openings in the roof and sides of a greenhouse located behind a mobile home. A majority of the justices in *Riley* found the observations constitutional.[16]

The Court's 2001 *Kyllo v. United States*[17] decision, though it considered ground-based observations of a home, is also relevant to UAS privacy. In 1992, a government agent in a car used a thermal imaging device to measure the external temperature of the roof and outside wall of the home of a person suspected of growing marijuana. The roof and wall were found to be abnormally warm, and a search warrant was issued based in part on this information. Upon execution of the search warrant, marijuana plants were found. In a decision finding the use of the thermal imager unconstitutional, the Court held that when "the Government uses a device that is not in general public use, to explore details of the home that would previously have been unknowable without physical intrusion, the surveillance is a 'search' and is presumptively unreasonable without a warrant."[18]

The most recent Supreme Court ruling with significant implications for UAS privacy is *United States v. Jones.*[19] *Jones* considered the government's installation, without a valid warrant, of a GPS tracking device on a vehicle used by a suspect in a narcotics investigation. The Court's January 2012 decision was unanimous in finding the government's actions unconstitutional, but there was considerable divergence in the basis for that finding. The majority opinion, delivered by Justice Scalia, found a Fourth Amendment violation in the physical trespass that occurred during the placement of the GPS device on the vehicle. That intrusion, wrote Justice Scalia, "would have been considered a 'search' within the meaning of the Fourth Amendment when it was adopted."[20]

In an opinion concurring in the judgment, Justice Alito criticized "the Court's reliance on the law of trespass" to decide the case.[21] Instead, he wrote, the question is whether the "respondent's reasonable expectations of privacy were violated by the long-term monitoring" of his vehicle.[22] Because "law enforcement agents tracked every movement that respondent made in the vehicle he was driving" for four weeks—a level of monitoring that Justice Alito felt impinged on reasonable expectations of privacy—Justice Alito concluded that the tracking constituted a search.[23]

Justice Sotomayor, in addition to joining the majority, provided a separate concurring opinion arguing that "the trespassory test . . . reflects an irreducible constitutional minimum"[24] and agreeing with Justice Alito's view that the respondent's reasonable expectations of privacy were violated. Justice Sotomayor also expressed concern that the unchecked ability of the government to assemble "the sum of one's public movements" could enable it to obtain private information regarding political and religious beliefs.[25]

A "Level of Detail" Test for Constitutionality of UAS Images?

In *Ciraolo* and *Riley*, the aerial observations of the curtilage of a home were made with the naked eye, a fact specifically noted by the *Ciraolo* Court and in both of the *Riley* opinions that found the government's observations constitutional. Writing for the plurality in *Riley*, Justice White found no Fourth Amendment violation in part because "no intimate details connected with the use of the home or curtilage were observed."[26]

In combination, the opinions in *Ciraolo* and *Riley* imply a level of detail threshold beyond which government UAS observations could cross the line into a Fourth Amendment search and thus be unconstitutional without a warrant. This would pose no conflict with *Kyllo,* which stopped well short of *endorsing* the constitutionality of using technology in general public use. Under a "level of detail" standard, even widely available UAS technology would lead to a Fourth Amendment violation if used by the government to capture intimate details of a home or its curtilage not normally discernable to a passer-by (or to a person flying over the house in a manned aircraft). By contrast, an image from a UAS camera flown at a very high altitude might raise no constitutional concerns if it reveals very few details.

LONG-TERM UAS SURVEILLANCE OF A HOME OR CURTILAGE

For logistical, regulatory, and financial reasons, law enforcement agencies will usually use small UAS. These platforms typically have flight durations measured in dozens of minutes. Thus, the overwhelming majority of law enforcement—and more generally, government—UAS will be incapable of conducting surveillance over a continuous period of days or weeks. But a full consideration of UAS privacy requires acknowledging that some unmanned aircraft do have long flight durations, and could potentially be used by the government to conduct extended surveillance of a home or its curtilage. Would this constitute a Fourth Amendment search?

Lower court rulings related to video cameras installed by the government on fixed structures such as utility poles to surveil private property in the course of criminal investigations are somewhat divergent, but generally recognize the higher expectation of privacy accorded the interior of a home and the fenced-in portion of curtilage.[27] In 1987, for example, the Fifth Circuit ruled on a case involving a government-installed video camera on a light pole overlooking a 10-foot fence at the border of a defendant's backyard. In finding this to be a Fourth Amendment search, the Fifth Circuit wrote that "[h]ere, unlike in *Ciraolo*, the government's intrusion is not minimal. It is not a one-time overhead flight or a glance over the fence by a passerby."[28] The court concluded that the defendant's "expectation to be free from this type of video surveillance in his backyard is one that society is willing to recognize as reasonable."[29]

A similar line of reasoning should apply to surveillance conducted using unmanned aircraft. The expectation to be free from from extended video surveillance in the curtilage of a home is eminently reasonable, and no less so if the monitoring is performed using a camera mounted on an unmanned aircraft as opposed to on a utility pole. This suggests that the Fourth Amendment will offer strong protection against long-term UAS surveillance of areas of a home or curtilage that a resident has sought to maintain as private.

LONG-TERM LOCATION TRACKING

Like multiple other technologies, unmanned aircraft could be used acquire images of pedestrian and vehicular traffic, and thus to potentially track people as they move about in public. In fact, the imaging technology to support this already exists. The DARPA-funded ARGUS-IS sensor merges data from an array of 368 imaging chips to form a single 1.8-gigapixel image. When used on an unmanned aircraft from a height of approximately 17,000 feet, ARGUS can take in a field of view covering many square miles while still retaining sufficient resolution to see six-inch objects on the ground.[30] Completely independently, Boeing is under contract with DARPA to develop the SolarEagle, a solar-powered UAS that will be able to stay aloft in the stratosphere for five continuous years.[31] Mounting an imaging sensor such as ARGUS on a UAS platform like the SolarEagle would create a system that could track all vehicle and pedestrian movements over many square miles on an extended basis.

The prospect that the government might someday be able to use unmanned aircraft to capture—and then use at will—this level of information about the movements of private individuals is sobering. It is therefore even more sobering that very similar tracking is *already* being conducted using the increasingly extensive networks of ground-level cameras and license plate readers deployed in many American cities. These systems collect information that, when aggregated, can track our travels just as thoroughly as a bank of cameras on a solar-powered unmanned aircraft turning slow circles in the stratosphere.

The existence of multiple technologies that can perform extended warrantless location tracking in no way diminishes the privacy concerns that would be raised when any one technology is used for this purpose. But the underlying constitutional question raised by location tracking by the government is not technology specific: Do we have a reasonable expectation of privacy from government monitoring of the *totality* of our public movements over extended periods? Most people would answer this question in the affirmative, and the opinions in *Jones* suggest that a majority of the Supreme Court Justices would answer affirmatively as well.

For the *Jones* majority, the physical trespass involved in attaching the tracking device was sufficient to find the government's actions unconstitutional. Justice Scalia's majority opinion noted that while the Court had previously found traditional visual surveillance using a large team of agents to be constitutional, extended location tracking using technologies such as GPS may be different: "It may be that achieving the same result through electronic means, without an accompanying trespass, is an unconstitutional invasion of privacy, but the present case does not require us to answer that question."[32]

However, five of the Justices wrote or joined concurrences that went further. In a concurrence joined by three other Justices, Justice Alito expressed a view that extended warrantless electronic surveillance of public movements is unconstitutional, writing, "the use of longer term GPS monitoring in investigations of most offenses impinges on expectations of privacy."[33] In a separate concurrence, Justice Sotomayor noted her agreement with this statement.[34] Notably, Justice Alito's classification of the tracking as a search was a direct consequence of its extended length:

> In this case, for four weeks, law enforcement agents tracked every movement that respondent made in the vehicle he was driving. We need not identify with precision the point at which the tracking of this vehicle became a search, for the line was surely crossed before the 4-week mark.[35]

We will have to wait for a future case to know whether the Court itself would adopt such a view as the basis for finding a search unconstitutional. But in the aggregate, the *Jones* opinions indicate that the Fourth Amendment will indeed provide protection against extended warrantless location tracking using technologies such as GPS or unmanned aircraft.

NON-GOVERNMENT UNMANNED AIRCRAFT AND PRIVACY

Private entities are not bound by Fourth Amendment restrictions that apply to the government, and have an affirmative First Amendment privilege to gather information. However, while that privilege is extensive, it is not unbounded, and can end when it crosses into an invasion of privacy. While most companies and individuals will endeavor to use unmanned aircraft responsibly, there will inevitably be some operators who observe less restraint. For example, it would be optimistic in the extreme to expect paparazzi to always operate UAS in a manner respectful of privacy considerations.

Invading an individual's privacy using a UAS could result in civil or criminal liability. Courts in most jurisdictions recognize two forms of common law invasion of privacy that could arise in connection with UAS: intrusion upon seclusion, and public disclosure of private facts. In addition, many states have civil or criminal statutes, or both, related to invasion of privacy. A person who is photographed in his or her own home by a UAS hovering just outside an otherwise inaccessible window would thus have strong grounds for a valid cause of action.

Although privacy expectations are greatly reduced outside the home, use of an unmanned aircraft to capture images and other information in a public setting could sometimes constitute an invasion of privacy. A 1998 California Supreme Court ruling in *Shulman v. Group W Productions, Inc.*[36] regarding filming that occurred following a car accident is instructive in this regard. In *Shulman* the victims of the accident were subjected to video and audio recording (using a traditional news camera held by a camera operator on the ground, combined with audio from a microphone worn by the flight nurse from a helicopter rescue crew) without their consent for a television program. The California Supreme Court found that a woman injured in the accident "was entitled to a degree of privacy in her conversations with [the nurse] and other medical rescuers at the accident scene."[37] It was improper, the ruling stated, to conclude that the "plaintiffs had no reasonable expectation of privacy at the accident scene itself because the scene was within the sight and hearing of members of the public."[38] At the time of the rescue, the vehicle was located "in a ditch many yards from and below" a highway, rendering it "unlikely . . . that any passersby on the road could have heard" the conversation of the victim with the nurse and other rescuers.[39]

The *Shulman* ruling provides important guidance regarding newsgathering and UAS with respect to intrusion upon seclusion by drawing a distinction between the information about a newsworthy event available to passersby and the potentially greater amount of information available only to those intimately involved in the event itself. Under some circumstances, there will be no difference: If a crowd of people has gathered around an assault victim who is being treated by paramedics, the conversation between the victim and the paramedics may well be audible to all, and the victim's reasonable expectation of privacy correspondingly lower. But there will be many cases in which, as in *Shulman*, passersby have far from complete information about an unfolding event. Using an unmanned aircraft to fill in the gaps would carry a risk of crossing the line into intrusion.

There is an opportunity to strengthen state invasion of privacy statutes to add UAS to the list of recited technologies that could be used to violate privacy. For example, a bill currently pending in the California State Legislature would amend the California civil code statute relating to constructive invasion of privacy to identify a "device affixed to or contained within an unmanned aircraft system" as one of the visual or auditory enhancing devices that could be used to violate a plaintiff's reasonable expectation of privacy.[40] The same bill would also amend the California penal code to explicitly provide that eavesdropping upon or recording a confidential communication using "a device affixed to or contained within" an unmanned aircraft is punishable by a fine or imprisonment.

BUSINESS PRIVACY

In 2011, a Texas man flew an unmanned aircraft over land near a Dallas-area meat packing plant and acquired images appearing to show environmental violations.[41] He contacted the Coast Guard, and in early 2012 the Texas Environmental Crimes Task Force served a search warrant on the company.[42] Once on the property, investigators found a pipe that "originate[d] in the back of the slaughterhouse" that appeared to be channeling pigs' blood into a nearby river and "[was] not linked to a waste water system."[43]

Though few would rush to defend a slaughterhouse that may be in violation of environmental codes, the ability of private citizens or groups to easily inspect for such violations via overflights raises complex issues. To the extent that overflights are lawfully conducted and reveal activities that may be endangering public health, they are obviously valuable. But what happens if a well-meaning but overzealous environmental group conducts daily flights over a large, fenced-in manufacturing facility and repeatedly reports "violations" to the government that turn out, after costly and time-consuming on-the-ground inspections, not to be violations after all? Does the company that owns the facility have grounds to ask a court to enjoin the environmental group from further overflights? What if the group posts pictures from its daily aerial surveillance missions on the Internet, and in doing so exposes information that the company regards as a trade secret? What if the group were to use an unmanned aircraft to examine the interior of buildings at the facility using an advanced thermal imager that, if in government hands and used to inspect a home, would be unconstitutional under *Kyllo*?

In many respects these questions involve issues that go well beyond unmanned aircraft. And in most cases, they will best be resolved through the courts as opposed to through new legislation aiming to solve a business UAS privacy problem of still-uncertain scope and nature.

CONCLUSION

The Founders of this country certainly did not have unmanned aircraft in mind when drafting the Bill of Rights. But neither were they contemplating thermal imaging—a technology that the Fourth Amendment proved to be eminently capable of addressing hundreds of years later.

The Supreme Court's Fourth Amendment jurisprudence indicates that the government will indeed have some latitude in acquiring information using unmanned aircraft. But it also conveys the existence of important bounds. Use of an unmanned aircraft to resolve details of a home or its curtilage unavailable through naked-eye observations, or to conduct long-term observations of a home's curtilage, or to perform extended location tracking could be credibly challenged as unconstitutional without a warrant.

Of course, the only way to know with certainty how the Fourth Amendment will be interpreted with respect to unmanned aircraft observations is to wait for the courts—and ultimately, the Supreme Court—to rule. In the meantime, however, it would be a mistake to draft legislation driven by an assumption that the Fourth Amendment will offer *no* protection from UAS observations, no matter how invasive. A far better approach is to view privacy legislation for government UAS as complementary to a constitutionally protected right "to be secure . . . against unreasonable searches and seizures" that will still have force and meaning in a world in which unmanned aircraft are common.

There is also a legislative role—mostly at the state level—to provide enhanced privacy protection from UAS operated by non-government entities.

State statutes relating to trespassing, invasion of privacy, harassment, and stalking should be examined and updated as appropriate to address potential privacy abuses using UAS.

When considering potential new statutory privacy protections, it is helpful to keep in mind what has occurred with the Internet and mobile telephones, two technologies that are associated with privacy threats that are in some respects much more significant than those that will arise from unmanned aircraft. Usage of both the Internet and mobile phones grew as fast as their underlying technologies enabled. As a result, the public and legislative dialog regarding how best to address the privacy issues they raise has been conducted with a strong appreciation of their benefits. By contrast, while the privacy concerns associated with domestic UAS are real and deserving of attention, they are getting significant focus long before the potential benefits of the technology are widely recognized.

If, in 1995, comprehensive legislation to protect Internet privacy had been enacted, it would have utterly failed to anticipate the complexities that arose after the turn of the century with the growth of social networking and location-based wireless services. The Internet has proven useful and valuable in ways that were difficult to imagine over a decade and a half ago, and it has created privacy challenges that were equally difficult to imagine. Legislative initiatives in the mid-1990s to heavily regulate the Internet in the name of privacy would likely have impeded its growth while also failing to address the more complex privacy issues that arose years later.

This doesn't mean that there is no role for new privacy initiatives. But in developing these initiatives, it is important to recognize the near impossibility of predicting all of the ways that a rapidly developing technology can be used—for good or for ill—in future years. Some of the best privacy protection may in fact lie not in statutory text drafted with a keen eye on the latest innovations in unmanned aircraft technology, but instead in constitutional text drafted over two hundred years ago.

Thank you again for the opportunity to testify on this important topic.

End Notes

[1] The acronym "UAS" is also sometimes expanded to "unmanned aerial systems."

[2] John Villasenor, Observations From Above: Unmanned Aircraft Systems and Privacy, 36 HARV. J.L. & PUB. POL'Y 457 (2013).

[3] Bill Hannan, History in Models, Models in History, MODEL AVIATION, Dec. 1986, at 78, 79–81.

[4] Pub. L. No. 112-95, 126 Stat. 11 (2012).

[5] Id. §§ 331–336, 126 Stat. at 72–78.

[6] RQ-4 Global Hawk Factsheet, THE OFFICIAL WEB SITE OF THE U.S. AIR FORCE, http://www.af.mil/information/factsheets/factsheet.asp?id=13225 (last visited May 4, 2013).

[7] Press Release, AeroVironment Inc., AeroVironment Develops World's First Fully Operational Life-Size Hummingbird-Like Unmanned Aircraft for DARPA, Feb. 17, 2011, available at http://www.avinc.com/resources/press_release/aerovironment_develops_worlds_first_fully_operational_life-size_hummingbird.

[8] Press Release, Wyss Institute for Biologically Inspired Engineering at Harvard University, Robotic Insects make first controlled flight, May 2, 2013, available at http://wyss.harvard.edu/vi ewpressrelease/110/.

[9] Andrew Chuter, Solar UAV Lands After Record 2 Weeks Aloft, DEFENSENEWS, July 23, 2010, http://www.defensenews.com/article/20100723/DEFSECT01/ 7230304/Solar-UAV-Lands-After-Record-2-Weeks-Aloft.

[10] There was also a case, Dow Chem. Co. v. United States, 476 U.S. 227 (1986), that considered aerial photography using a mapping camera of the open areas of an industrial facility. The Supreme Court ruled that the open areas were more akin to an "open field" than to the curtilage of a home, and as a result, were "open to the view and observation of persons in aircraft lawfully in the public airspace immediately above or sufficiently near the area for the reach of cameras." Id. at 239.

[11] 476 U.S. 207 (1986).

[12] Ciraolo, 476 U.S. at 213.

[13] Id. at 213–14.

[14] Id. at 215.

[15] 488 U.S. 445 (1989).

[16] The Riley decision comprised an opinion delivered by Justice White and joined by three other Justices and an opinion from Justice O'Connor concurring in the judgment. Thus, while there was no majority opinion, a majority of the Justices found the observations constitutional.

[17] 533 U.S. 27 (2001).

[18] Id. at 40.

[19] 132 S. Ct. 945 (2012).

[20] Id. at 949.

[21] Id. at 962 (Alito, J., concurring).

[22] Id. at 958.

[23] Id. at 964.

[24] Id. at 955 (Sotomayor, J., concurring).

[25] Id. at 956.

[26] Riley, 488 U.S. at 452.

[27] See, e.g., United States v. Cuevas-Sanchez, 821 F.2d 248 (5th Cir. 1987) and United States v. Jackson, 213 F.3d 1269 (10th Cir. 2000).

[28] Cuevas-Sanchez, 821 F.2d at 251.

[29] Id.

[30] Ryan Gallagher, Could the Pentagon's 1.8 Gigapixel Drone Camera Be Used for Domestic Surveillance?, SLATE, Feb. 6, 2013, http://www.slate.com/blogs/future_tense/2013/02/06/argus_is_could_the_pentagon_s_1_8_gigapixel _drone_camera_be_used_for_domestic.html.

[31] Press Release, Boeing Co., Boeing Wins DARPA Vulture II Program (Sept. 15, 2010), available at http://boeing.mediaroom.com/index.php?s=43&item=1425.

[32] United States v. Jones, 132 S. Ct. 945, 954 (2012).

[33] Id. at 964 (Alito, J., concurring).

[34] Id. at 955 (Sotomayor, J., concurring).

[35] Id. at 964 (Alito, J., concurring).

[36] Shulman v. Group W Productions, Inc., 955 P.2d 469 (Cal. 1998).

[37] Id. at 491.

[38] Id.

[39] Id.

[40] S. B. 15, 2013–2014 Legis., Reg. Sess. (Cal. 2013), available at http://www.leginfo.ca.gov/pub/13-14/bill/sen/sb_0001- 0050/sb_15_bill_20130422_amended_sen_v97.htm.

[41] Meghan Keneally, Drone plane spots a river of blood flowing from the back of a Dallas meat packing plant, MAILONLINE, Jan. 24, 2012, http://www.dailymail.co.uk/news/article-2091159/A-drone-plane-spots-river-blood-flowing-Dallas-meat-packing-plant.html.

[42] Id.

[43] Id.

In: Drones in the U.S.
Editor: Sophia Barrett
ISBN: 978-1-63484-918-0

Chapter 5

TESTIMONY OF GREGORY S. MCNEAL, ASSOCIATE PROFESSOR OF LAW, PEPPERDINE UNIVERSITY SCHOOL OF LAW. HEARING ON "EYES IN THE SKY: THE DOMESTIC USE OF UNMANNED AERIAL SYSTEMS"*

The looming prospect of expanded use of unmanned aerial vehicles, colloquially known as drones, has raised understandable concerns regarding privacy. Those concerns have led some to call for legislation mandating that nearly all uses of drones be prohibited unless the government has first obtained a warrant. Such an approach would exceed the requirements of the Fourth Amendment and lead to perverse results that in some instances would prohibit the use of information when gathered by a drone, but would allow the same information to be admitted if gathered by nearly any other means. Such a technology centric approach to privacy misses the mark --- if privacy is the public policy concern, then legislation should address the gathering and use of information in a technology neutral fashion. This testimony outlines six key issues that Congress should remain cognizant of when drafting legislation.

---

* This is an edited, reformatted and augmented version of testimony presented May 17, 2013 before the House Judiciary Committee, Subcommittee on Crime, Terrorism, Homeland Security and Investigations.

1. Congress Should Reject Calls for a Blanket Requirement That All Drone Use Be Accompanied by a Warrant

Proposals that prohibit the use of drones for the collection of evidence or information unless authorized by a warrant are overbroad and ill-advised.[1] Such legislation treats the information from a drone differently than information gathered from a manned aircraft, differently than that gathered by a police officer in a patrol car, or even from an officer on foot patrol. Under current Fourth Amendment jurisprudence, police are not required to shield their eyes from wrongdoing until they have a warrant, why impose such a requirement on the collection of information by drones?

For example, imagine a police officer was on patrol in her patrol car. While driving she witnesses the car in front of her strike a pedestrian and speed off. Until witnessing the crime she did not have probable cause, or even reasonable suspicion to believe the vehicle in front of her would be involved in a crime. Let's further assume that her dash camera recorded the entire incident. That video may be used as evidence against the driver in a subsequent criminal proceeding, but under broadly worded proposals mandating a warrant for drone usage, the same piece of evidence if gathered by a drone would be inadmissible in court. Why?

Consider another example. Police receive an anonymous tip that someone is growing marijuana in their backyard. A police officer attempts to view the backyard from the ground but his view is blocked by a 10 foot tall fence. The officer next decides to fly a commercially available remote controlled helicopter[2] over the backyard and from a vantage point that does not violate FAA regulations observes marijuana plants growing in the yard. This observation would be unlawful under proposals that require a warrant for observations from a drone. However, these facts are nearly identical to the facts in the Supreme Court's 1986 *California v. Ciraolo*[3] decision which upheld aerial surveillance. The only difference is that in *Ciraolo*, the officer flew over the backyard in an airplane, rather than using a drone.

Notably, the fact that Ciraolo had erected a 10 foot fence to manifest his "intent and desire" to maintain privacy did not necessarily demonstrate a reasonable expectation of privacy as the court noted that the fence "might not shield these plants from the eyes of a citizen or a policeman perched on the top of a truck..." Thus, according to the Supreme Court, not only would observation of the marijuana plants from the air (as described above) be lawful, observation from the top of a police truck over the fence would be lawful, and by extension, observation of the marijuana plants by police from the third floor of a neighboring home would also be lawful. But under proposals requiring a warrant for observations by a drone, this evidence would be inadmissible.

What public policy goal is advanced by the suppression of evidence of a crime when documented by a drone when the same evidence if recorded by a dashcam, observed from an airplane, or viewed from a neighboring home would be admissible in court?

2. Congress Should Reject Broadly Worded Use Restrictions

Congress should reject broadly worded use restrictions that prohibit the use of any evidence gathered by drones in nearly any proceeding. Such restrictions exceed the parameters of the Fourth Amendment and in some circumstances may only serve to protect criminals while not deterring governmental wrongdoing.

For example, the Alameda County, California Sheriff's Department recently proposed the use of small UASs for: crime scene documentation, EOD missions, HAZMAT response, search and rescue, public safety and life preservation missions, disaster response, fire prevention, and documentation of a felony when such documentation is premised upon probable cause.[4] Linda Lyle, a privacy advocate with the ACLU criticized the proposal, stating: "If the sheriff wants a drone for search and rescue then the policy should say he can only use it for search and rescue...Unfortunately under his policy he can deploy a drone for search and rescue, but then use the data for untold other purposes. That is a huge loophole, it's an exception that swallows the rule."[5] Her points mirror the ACLU's position in their December 2011 white paper where they state that drone use is acceptable so long as "the surveillance will not be used for secondary law enforcement purposes."[6] It is also similar to the language used in other proposals prohibiting the use of information gathered by a drone "as evidence against an individual in any trial, hearing or other proceeding...."

A simple hypothetical can help to illustrate the problem with this approach. Imagine that law enforcement uses a drone to search for a lost hiker in a national park. This is a search and rescue mission that fits within the public safety, emergency, or exigency exceptions in most proposals. However, imagine that during the course of the search the drone observed a man stabbing a woman to death in the park. That collection was entirely inadvertent, and as such suppressing the videotape of the stabbing would not serve to deter the police from using drones in the future as they were not searching for an unrelated stabbing crime, they were searching for a lost hiker. Yet, that evidence under the blanket use restrictions found in various proposals circulating in state legislatures, Congress, and under the ACLU's "secondary law enforcement purposes" standard would need to be suppressed.[7] Such suppression doesn't protect privacy (as inadvertent discovery can't be deterred); it merely protects a criminal who if observed from a helicopter, an airplane, or from the ground would face evidence of his crime, but under some broadly worded drone focused privacy bills may be more difficult to prosecute.

What public policy goal is furthered by suppressing evidence of a crime merely because the evidence was gathered from a drone? If the discovery were genuinely inadvertent, there is little to no deterrent value that justifies suppressing the evidence.

3. If Congress Chooses to Impose a Warrant Requirement, It Should Carefuly Consider Codifying Exceptions

If Congress seeks to impose a statutory warrant requirement on the use of drones, it should codify exceptions to the warrant requirement and exclusionary rule that the courts have developed through decades of jurisprudence.

As the Supreme Court has noted, suppressing evidence has serious consequences for the "truth-seeking and law enforcement objectives" of our criminal justice system, and as such should present "a high obstacle for those urging [for its] application"[8] it should be "our last resort, not our first impulse."[9] As such, the measure for when we should apply the exclusionary rule should not be whether a drone was used, but rather should be when "the benefits of deterrence...outweigh the costs."[10] Some exceptions and other procedural devices that Congress should consider codifying are:

- Rather than codify a blanket restriction on the use of any information gathered from a drone, Congress should codify a *standing requirement* that premises one's ability to raise a suppression challenge on whether the person raising the suppression claim was the purported target of drone surveillance. Thus, if law enforcement uses a drone to document illegal dumping of toxic waste by Co-conspirator #1, Co-conspirator #2's privacy rights were not violated and #2 should not have the ability to vicariously assert #1's privacy rights to protect himself from prosecution.
- Evidence gathered by drones should be admissible in proceedings short of trial such as grand jury proceedings,[11] preliminary hearings,[12] bail hearings,[13] and other *non-trial proceedings.*
- Evidence gathered by drones should be admissible for *impeachment purposes* as there is little deterrent value in keeping such impeachment evidence out of a trial (as law enforcement is unlikely to gather it solely for that purpose) and the use of evidence gathered by drones for such a limited purpose furthers the truth-seeking process and deters perjury.[14]
- If Congress imposes a statutory warrant requirement on the use of drones, it should also codify directly, or by reference the body of jurisprudence associated with the so-called *good faith exception* as articulated in *United States v. Leon*[15] and *Massachusetts v. Sheppard*.[16] The good faith exception allows for the admission of evidence gathered pursuant to a defective warrant, unless, based on objective facts, "a reasonably well trained officer would have known the search was illegal despite the magistrate's authorization."
- Congress should make clear that the *independent source doctrine* as articulated in *Murray v. United States* applies equally to drone related surveillance.[17] The independent source doctrine allows for the admission of evidence, despite police illegality, if the evidence seized was not causally linked to the illegal police conduct.
- Congress should codify the *inevitable discovery rule* articulated in *Nix v. Williams*.[18] In the context of drone surveillance, the rule would operate to allow the admission of drone gathered evidence in a criminal trial if the prosecutor can prove (by a preponderance of the evidence) that the evidence would have ultimately or inevitably been discovered by lawful means.[19]

- Rather than suppress all fruit of drone surveillance, Congress should codify the *attenuation principles* articulated in *Nardone* and *Wong Sun*.[20] The Court in *Wong Sun* stated that when considering whether fruit of an unlawful search should be suppressed, a court must ask "whether, granting establishment of the primary illegality, the evidence to which instant objection is made has been come at by exploitation of that illegality or instead by means sufficiently distinguishable to be purged of the primary taint." Stated differently, at some point the fruit of the poisonous tree loses its potency. Factors Congress should consider codifying are passage of time between the illegal search and the acquisition of evidence, intervening events and a lack of foreseeability that the illegal drone surveillance would result in the gathering of evidence, and whether the initial illegal surveillance was a flagrant or deliberate violation rather than an accidental one.[21]

4. Congress Should Spend a Substantial Amount of Time Carefully Defining Terminology and Specifying What Places Are Entitled to Privacy Protection

What a layperson sees when they read the word search, what a legislator means, and what a court may think the legislature meant are all different things. As such, when using terms like search, surveillance, reasonable expectations, curtilage, private property, public place and other terms of art, Congress should specify what the terms mean. This definitional task will be the most important part of the legislative drafting process as the terminology will drive what actions are allowable and what places are entitled to privacy protection. Congress should consider adopting an entirely new set of definitions, and be prepared to reject existing terminology which may be confusing. A good example of a well thought out definitional approach is the proposed legislation offered by Professor Christopher Slobogin.[22] He uses the following terms:

- "Search: An effort by government to find or discern evidence of unlawful conduct. A targeted search seeks to obtain information about a specific person or circumscribed place. A general search seeks to obtain information about people or places that are not targets at the time of the search."

- "Public search: A search of a place, in the absence of explicit consent, focused on activities or persons, limited to what the natural senses of a person on a lawful public vantage point could discern at the time of the search.
- "Probable cause: An articulable belief that a search will more likely than not produce contraband, fruit of crime, or other significant evidence of wrongdoing..."
- "Reasonable suspicion: An articulable belief that a search will more likely than not lead to evidence of wrongdoing...."

5. Congress May Want to Consider Crafting Simple Surveillance Legislation, Rather Than Very Detailed Drone Legislation

In light of the various issues I've raised in my prior points, Congress may find it preferable to legislate with an eye towards controlling surveillance writ large, not just drone surveillance. To do this Congress should focus on controlling the duration of surveillance.

The duration of surveillance can be controlled by crafting legislation that places aggregate limits on how long law enforcement may surveil specific persons or places. Slobogin suggests a sliding scale, allowing for 20 minute searches at an officer's discretion, 20 minute to 48 hour searches with a court order and reasonable suspicion, and searches of longer than 48 hours when accompanied by a warrant and probable cause.[23] The specific amount of time Congress may settle on will depend on whether Congress wants to value privacy or law enforcement efficiency, but the point is that carefully crafting duration based rules for surveillance (whether by drone or otherwise) may be a wiser choice than the current drone focused approach that is riddled with blanket bans and exceptions. (To see the perils of a process riddled with exceptions, look at the recent bill passed by the Texas legislature which has no fewer than 22 exceptions for drone use with carve outs for agriculture interests, electrical companies, oil companies, real estate brokers and others).[24] Rather than crafting special exceptions, legislating with an eye towards creating rules based on clearly defined (albeit arbitrary) durational limits on surveillance creates legislation that is clearer and easier to follow.

6. Congress Should Consider Adopting Transparency and Accountability Measures, Perhaps in Lieu of a Warrant Requirement or Suppression Rules

Transparency and accountability measures may be more effective than suppression rules or warrants for controlling and deterring wrongful government surveillance. To hold law enforcement accountable, Congress should mandate that the use of all drones be published on a regular basis (perhaps quarterly) on the website of the agency operating the system. These usage logs should detail who operated the system, when it was operated, where it was operated (including GPS coordinates), and what the law enforcement purpose for the operation was. Congress may even mandate that manufacturers of unmanned systems come equipped with software that allows for the easy export of flight logs that contain this information. Such logs will allow privacy advocates and concerned citizens to closely monitor how drones are being used and enables the political process as a mechanism to hold operators accountable.

In circumstances where publishing usage logs may reveal information that is law enforcement sensitive (such as an ongoing investigation) the agency operating the drone may keep their usage logs confidential until the investigation is closed. The agency should be required to make the logs public within 30 days of the close of an investigation. To facilitate public accountability Congress should mandate that all logs be published in an open and machine readable format consistent with the President's Executive Order of May 9, 2013.

For evidence that this flight log approach works, one need only look across the Atlantic to the UK where many police departments publish their helicopter flight logs on their webpage --- in fact some even live Tweet their helicopter's activities. While there is no law within the United Kingdom that specifically requires police departments or law enforcement agencies to publish the flight logs of their helicopters, their version of the Freedom of Information Act appears to be the legislative authority prompting publication of police helicopter logs.

Like the United States, there are a number of public watchdog groups in the United Kingdom that monitor police activity, including groups whose sole purpose is to monitor the activity (and related noise complaints) of police helicopters.[25] These groups, and their respective websites, act as a forum for noise and privacy complaints from various individuals across the Kingdom, and several of these groups organize and lobby Members of Parliament (MPs) to pass legislation restricting helicopter flyovers. [26] These groups, and the advocacy which they generate, appear to be largely responsible for the recent trend of many UK police departments publishing their helicopters' flight logs, or even creating Twitter accounts for their helicopters that publish real-time or delayed-time updates of the aircraft's activity.[27]

These helicopter Twitter accounts, which have become a growing trend amongst British police departments, have had an immediate and powerful effect on public relations in their respective jurisdictions. In Islington, the police department went from struggling to handle the overload of noise complaints relating to the department's use of its helicopter to receiving no complaints after the creation of its Helicopter Twitter feed.[28] The Twitter account gained over 7,000 followers after its first few weeks, and the public criticism of police helicopter activity ceased entirely. The officer second in command of the department reflected on the effectiveness—as well as future potential— of the Twitter feed by issuing this statement:

> Maybe that is all people wanted – just to know and understand what we were doing. We don't update people in real time, but my vision is that soon we will be able to let people know about an operation as soon as it is over. In some cases we could get them to help – imagine if an elderly person with Alzheimer's was missing in Islington, we could Tweet our followers to keep an eye out.

The Suffolk Police Department launched its Twitter feed with the hope of shedding some light on police practices. Roger Lewis, an observer with the Suffolk Police, described the department's intentions in the following way:

> We hope to use the Twitter feed to highlight the positive work being done by the Air Operations Unit and to keep members of the public informed as to why the helicopter has been deployed. We hope people will enjoy finding out more about the Unit and hopefully our tweets will give some explanation as to why we have been deployed and give some interesting insights into a very important policing tool.[29]

It is not difficult to see how the practice of disclosing non-sensitive flight logs through a public channel—such as a department web page or through Twitter—can be a useful tool in reassuring the public that law enforcement's helicopter does not represent Big Brother's eye in the sky, but rather embodies a part of the department's lawful policing practices. Just as a police helicopter high overhead can be ominous to those on the ground who are unaware of its purposes, the very idea of drones—of any kind—flying above American cities and towns might be foreboding to many lay persons. By requiring law enforcement to publish data or logs Congress can add a citizen centric political check that will help quell the fears of a society that is not yet certain how it should react to the increasing presence of drones over the skies of America.

CONCLUSION

The emergence of unmanned aerial vehicles in domestic skies raises understandable privacy concerns that require careful and sometimes creative solutions. However, we should reject alarmist calls that suggest we are on the verge of an Orwellian police state as we've heard these calls before and they did not come true.[30] In 1985, the ACLU argued in an amicus brief filed in *California v. Ciraolo* that police observation from an airplane was "invasive modern technology" and upholding the search of Ciraolo's yard would "alter society's very concept of privacy." Later, in 1988, the ACLU argued in *Florida v. Riley* that allowing police surveillance by helicopter was "Orwellian" and "would expose all Americans, their homes and effects, to highly intrusive snooping by government agents..." In a different context in 2004 (before the advent of the iPhone) police in Boston were going to use Blackberry phones to access public databases --- the equivalent of Googling. Privacy advocates decried the use of these handheld phones as "mass scrutiny of the lives and activities of innocent people," and "a violation of the core democratic principle that the government should not be permitted to violate a person's privacy, unless it has a reason to believe that he or she is involved in wrongdoing."[31] Reactionary claims such as these get the public's attention and are easy to make, but have the predicted harms come true? Is the sky truly falling? We should be careful to not craft hasty legislation based on emotionally charged rhetoric. Outright bans on the use of drones and broadly worded warrant requirements that function as the equivalent of an outright ban do little to protect privacy or public safety and in some instances will only serve to protect criminal wrongdoing. Rather than pursuing a drone specific approach or a warrant based approach, Congress should consider surveillance legislation aimed at making the use of these systems more transparent and empowering the people to hold operators accountable.

End Notes

[1] For example, the "Preserving Freedom from Unwarranted Surveillance Act of 2012."

[2] Perhaps a Parrot A.R. drone from the local mall's Brookstone store.

[3] 476 U.S. 207 (1986).

[4] Alameda County Sheriff's Office, General Order 615 available at: http://nomby.files.wordpress.com/2013/02/smallunmanned-aircraft-system-general-order-6-15-draft.pdf

[5] Paul Detrick, "Cops with Drones: Alameda Co. CA Weighs Technology vs. Privacy" available at: http://reason.com/reasontv/2013/04/04/cops-with-drones-technology-vs-privacy

[6] "Protecting Privacy from Aerial Surveillance: Recommendations for Government Use of Drone Aircraft," American Civil Liberties Union, December 2011, p. 16.

[7] For example, the "Preserving American Privacy Act of 2013" in Section 3119c creates a general prohibition on the use of covered information as "evidence against an individual in any trial, hearing, or other proceeding..." While the Act provides a set of exceptions, including one for emergencies, the language of the emergency exception as currently drafted does not clearly specify that inadvertent discovery of information unrelated to the emergency justifying the drone usage would be admissible, and it's likely that defense counsel in such a case would seek to prohibit the admission of evidence in such a case by relying on the lack of a clearly specified exception.

[8] Pennsylvania Board of Probation and Parole v. Scott, 524 U.S. 357, 364-365 (1998).

[9] Hudson v. Michigan, 547 U.S. 586, 591 (2006).

[10] Herring v. United States, 129 S. Ct. 695, 700 (2009).

[11] This is consistent with the Supreme Court's approach to Fourth Amendment violations per United States v. Calandra, 414 U.S. 338 (1974). (noting that allowing "a grand jury witness to invoke the exclusionary rule would unduly interfere with the effective and expeditious discharge of the grand jury's duties, and extending the rule to grand jury proceedings would achieve only a speculative and minimal advance in deterring police misconduct at the expense of substantially impeding the grand jury's role.").

[12] This is consistent with Congress' guidance in Federal Rule of Criminal Procedure 5.1(e) which states in relevant part "At the preliminary hearing, the defendant may cross-examine adverse witnesses and may introduce evidence but may not object to evidence on the ground that it was unlawfully acquired."

[13] See 18 U.S.C. 3142(f) noting the "rules concerning admissibility of evidence in criminal trials do not apply to the presentation and consideration of information at the hearing."

[14] Contra James v. Illinois, 493 U.S. 307 (1990).

[15] 468 U.S. 897 (1984).

[16] 468 U.S. 981 (1984).

[17] 487 U.S. 533 (1988).

[18] 467 U.S. 431 (1984).

[19] Note, Nix was a Sixth Amendment case but courts have applied the fruits analysis to searches.

[20] Nardone v. United States, 308 U.S. 338 (1939) and Wong Sun v. United States 371 U.S. 471 (1963) respectively.

[21] See, Brown v. Illinois, 422 U.S. 590, 604 (1975).

[22] Christopher Slobogin, "Making the Most of United States v. Jones in a Surveillance Society: A Statutory Implementation of Mosaic Theory," Duke Journal of Constitutional Law & Public Policy (forthcoming) available at: http://papers.ssrn.com/sol3/papers.cfm?abstract id=2098002 .

[23] Slobogin at 24.

[24] See HB 912, available at: http://www.legis.state.tx.us/BillLookup/History .aspx?LegSess= 83R&Bill=HB912.

[25] See http://www.helicopter-noise.org.uk/; http://www.whatdotheyknow.com/request/issue of police helicopter fligh

[26] See http://www.parliament.uk/edm/2012-13/394 (proposed legislation to regulate/reduce the amount of noise pollution caused by nighttime police helicopter flyovers in London).

[27] Not all activity is published. The Cleveland (UK) Police Department's website indicates that: "This page is intended to provide basic information to the general public regarding the work of the police helicopter and will be updated on a daily basis. Weekend and public holiday updates will appear on the next working day. Please note that not all items are always listed due to operational sensitivity or ongoing investigation." http://www.cleveland. police.uk/news/helicopterwatch.aspx

[28] http://www.islingtongazette.co.uk/news/police helicopter twitter account stops islington complaints 1 1206725

[29] http://helihub.com/2012/09/03/uks-suffolk-police-helicopter-unit-now-on-twitter/

[30] Interestingly, Orwell seems to be a favorite citation for the ACLU who has cited him nearly 70 times in briefs.

[31] See Gregory S. McNeal, "Can The 'Drone' Industry Compete With The Privacy Lobby?" available at: http://www.forbes.com/sites/gregorymcneal/2012/08/13/can-the-drone-industry -compete-with-the-privacy-lobby/

In: Drones in the U.S.
Editor: Sophia Barrett
ISBN: 978-1-63484-918-0

Chapter 6

STATEMENT OF TRACEY MACLIN, PROFESSOR OF LAW, BOSTON UNIVERSITY SCHOOL OF LAW. HEARING ON "EYES IN THE SKY: THE DOMESTIC USE OF UNMANNED AERIAL SYSTEMS"*

Thank you Chairman Sensenbrenner, ranking Member Scott, and members of the Committee for inviting me to testify about the Fourth Amendment issues surrounding the domestic use of drones by law enforcement officials.

My name is Tracey Maclin. I am a law professor at Boston University School of Law. I have taught Constitutional Law and Criminal Procedure since 1985. My opening comments will be brief.

The constitutionality of drones for domestic law enforcement purposes raises several questions that are not easily answered by the Supreme Court's current jurisprudence.

Drones can be equipped with sophisticated cameras, thermal imaging devices, license plate readers and laser radar systems. According to a recent paper by the Congressional Research Service, soon drones will be able to operate with facial recognition or soft biometric recognition equipment that can "recognize and track individuals based on attributes such as height, age, gender, and skin color."

---

* This is an edited, reformatted and augmented version of a statement presented May 17, 2013 before the House Judiciary Committee, Subcommittee on Crime, Terrorism, Homeland Security and Investigations.

Because of the advanced technology now available, comparing a drone to a traditional airplane for Fourth Amendment purposes is similar to comparing a frisk conducted by a security guard to a modern x-ray machine that can see beneath one's clothing utilized at some airports.

The 1980s Supreme Court rulings that airplane and helicopter surveillance did not implicate the Fourth Amendment were premised on naked-eye searches and surveillance equipment that was readily available to the public. For example, in *California v. Ciraolo*, Chief Justice Burger's majority opinion distinguished concerns about future electronic developments from "*simple visual observations from a public place*" that were challenged in *Ciraolo*.

Moreover, in each of those cases, the Court signaled that more intrusive and sophisticated police surveillance would raise different and more difficult Fourth Amendment issues. Thus, I agree with the view of John Villasenor that the Court's 1980's rulings do not control the use of drones that are capable of capturing much more detail unavailable to the human eye.

Furthermore, it is important to recognize, even among the Justices of the Roberts' Court, the definition of what constitutes a "search" under the Fourth Amendment is subject to change.

In the recent GPS case, *United States v. Jones*, five Justices indicated a willingness to reassess traditional notions of privacy regarding long-term electronic monitoring of a person's movements in public. Justice Sotomayor, for example, encouraged her colleagues to reconsider the Court's traditional analysis for even short-term monitoring of a person's public activities. And Justice Alito, although not going as far as Justice Sotomayor in his willingness to reconsider the Court's privacy jurisprudence, did state: "The use of longer term GPS monitoring investigations of most offenses impinges on expectations of privacy."

A final point: When considering whether drone surveillance constitutes a search under the Fourth Amendment, I encourage the Committee to avoid resolving this question with a simplistic litmus test or legal term of art.

When deciding search and seizure cases, often judges will rule that the central point of the Fourth Amendment is "reasonableness." In the typical case, this "reasonableness" model is the equivalent of a rational basis test for judging a statute, or governmental conduct that implicates non-fundamental rights. In these ordinary cases, judges uphold police intrusions because the intrusion rationally serves legitimate state interests.

This degree of deference to police intrusions is at odds with the central purpose of the Fourth Amendment, which is distrust of discretionary police power. The Fourth Amendment was not inserted in the Bill of Rights so that judges could defer to government intrusions of privacy; rather the amendment was designed to control such intrusions. The colonists who battled the British did not trust or defer to the judgments of British customs officials. They wanted the discretionary power of customs officers restrained.

Objections to warrantless drone surveillance do not stem from a view that law enforcement officers are bad people. Instead, objections are raised against this form of police authority because "power is a heady thing, and history shows that the police acting on their own cannot be trusted." This distrust of discretionary police power is the central meaning of the Fourth Amendment.

Thank you for the opportunity to speak today. I look forward to your questions.

In: Drones in the U.S.
Editor: Sophia Barrett
ISBN: 978-1-63484-918-0

Chapter 7

STATEMENT OF CHRISTOPHER R. CALABRESE, LEGISLATIVE COUNSEL, AMERICAN CIVIL LIBERTIES UNION. HEARING ON "EYES IN THE SKY: THE DOMESTIC USE OF UNMANNED AERIAL SYSTEMS"*

Good morning Chairman Sensenbrenner, Ranking Member Scott, and Members of the Committee. Thank you for the opportunity to testify on behalf of the American Civil Liberties Union (ACLU) its more than half a million members, countless additional activists and supporters, and fifty-three affiliates nationwide, about the privacy and civil liberties implications of the domestic use of unmanned surveillance vehicles, also known as drones, and recommendations for new protections for use of this technology.

I. INTRODUCTION

Unmanned aircraft carrying cameras raise the prospect of a significant new avenue for the surveillance of American life. Many Americans are familiar with these aircraft, commonly called drones, because of their use

* This is an edited, reformatted and augmented version of a statement presented May 17, 2013 before the House Judiciary Committee, Subcommittee on Crime, Terrorism, Homeland Security and Investigations.

overseas in places like Afghanistan and Yemen. But drones are coming to America. Recently passed legislation requires the Federal Aviation Administration to "develop a comprehensive plan to safely accelerate the integration of civil unmanned aircraft systems into the national airspace system."[1] This new legislation has dramatically accelerated the deployment of drones and pushed this issue to the forefront. Meanwhile, the technology is quickly becoming cheaper and more powerful, interest in deploying drones among police departments is increasing, and our privacy laws are not strong enough to ensure that the new technology will be used responsibly and consistently with constitutional values. In short, the specter of routine aerial surveillance in American life is on the near horizon — a development that would profoundly change the character of public life in the United States.

We need a system of rules to ensure that Americans can enjoy the benefits of this technology without bringing our country a large step closer to a "surveillance society" in which every move is monitored, tracked, recorded, and scrutinized by the authorities. This statement outlines a set of protections that would safeguard Americans' privacy in the coming world of drones.

Aerial surveillance from manned aircraft has been with us for decades. One of the first aircraft the Wright brothers built was a surveillance aircraft, and it was sold to the U.S. Army. Many common uses of drone aircraft—search and rescue, fighting wildfires, dangerous tactical police operations—are beneficial. In the 1980s the Supreme Court ruled that the Fourth Amendment does not categorically prohibit the government from carrying out warrantless aerial surveillance of private property.

But manned aircraft are expensive to purchase, operate and maintain, and this expense has always imposed a natural limit on the government's aerial surveillance capability. Now that surveillance can be carried out by unmanned aircraft, this natural limit is eroding. The prospect of cheap, small, portable flying video surveillance machines threatens to eradicate existing practical limits on aerial monitoring and allow for pervasive surveillance, police fishing expeditions, and abusive use of these tools in a way that could eventually eliminate the privacy Americans have traditionally enjoyed in their movements and activities. In order to prevent this harmful and invasive outcome, Congress must act.

II. THE TECHNOLOGY

There are hundreds of different types of Unmanned Aerial Vehicles (UAVs), as drones are formally known.[2] They can be as large as commercial aircraft or as small as hummingbirds, and include human remotely guided aircraft as well as autonomous, self-guided vehicles. They include:

- *Large fixed-wing aircraft.* The largest UAVs currently in use, such as the Israeli-made Eitan, are about the size of a Boeing 737 jetliner. The Eitan's wingspan is 86 feet, and it can stay aloft for 20 hours and reach an altitude of 40,000 feet.[3] The Predator B drone, which has been used extensively on overseas battlefields as well as on the U.S.-Mexico border, has a wingspan of 66 feet, and it can stay aloft for over 30 hours and reach an altitude of 50,000 feet.[4] In Pakistan and Afghanistan, the U.S. military and CIA deploy Predators and Reapers armed with surveillance capability as well as missiles capable of destroying a moving vehicle from thousands of feet in the air.[5]
- *Small fixed-wing aircraft.* Smaller fixed-wing aircraft are the current favorite for domestic deployment. The Houston police department, for example, recently tested the ScanEagle, made by Boeing subsidiary Insitu.[6] The ScanEagle is 4 1/2 feet long with a wingspan of 10 feet, and it can climb to 19,500 feet and stay aloft for more than 24 hours.[7]
- *Backpack craft.* Another class of craft is designed to be carried and operated by a single person. The hand-launched AeroVironment Raven, for example, weighs 4 pounds, has a wingspan of 4.5 feet and a length of 3 feet, can fly up to 14,000 feet and stay aloft for up to 110 minutes. Similar-sized products include a three-foot helicopter called the Draganflyer X6, a one-foot-long, one-pound fixed-wing craft called the AeroVironment Wasp, and a fan-propelled craft called the Honeywell T-Hawk that can "hover and stare." Individual hobbyists have also built a number of drones in this size range.[8]
- *Hummingbirds.* A tiny drone called the Nano Hummingbird was developed for the Pentagon's Defense Advanced Research Projects Agency (DARPA) by AeroVironment. Intended for stealth surveillance, it can fly up to 11 miles per hour and can hover, fly sideways, backwards and forwards, for about 8 minutes. It has a wingspan of 6.5 inches and weighs only 19 grams—less than a single AA battery.[9]

- *Blimps.* Some blimps are envisioned as high-altitude craft, up to 300 feet in diameter, that would compete with satellites, while others would be low-altitude craft that would allow the police to monitor the streets. Supporters say they are more cost-effective than other craft due to their ability to stay aloft for extended periods.[10]

III. DRONE CAPABILITIES—TODAY AND IN THE FUTURE

The aircraft themselves are steadily improving and, as with so many technologies, that is likely to continue. They are becoming smaller. The military and law enforcement are keenly interested in developing small drones, which have the advantages of being versatile, cheap to buy and maintain, and in some cases so small and quiet that they will escape notice.[11] They are also becoming cheaper. The amazing continual decreases in the prices of electronics that have become normal in our time all but guarantee that the surveillance technologies attached to UAVs will become less expensive and yet more powerful—and with mass production, the aircraft that carry those electronics will become inexpensive enough for a police department to fill the skies over a town with them.

Drones are also becoming smarter. Artificial intelligence advances will likely help drones carry out spying missions. Korean researchers, for example, are working to teach robots how to hide from and sneak up upon a subject.[12] They also will have better staying power, with a greater ability to stay aloft for longer periods of time. Mechanisms for increasing time aloft could include solar power, or the use of blimps or gliders.[13]

Although the primary uses of drones so far have been military, even on overseas battlefields their main use is surveillance. The larger drones can be fitted with weapons or other heavy payloads, but all of them can carry cameras and other imaging technologies that have developed amazing capabilities in recent years and are likely to become even more capable in the near future.

Except for possibly the very lightest craft, drones can carry the full range of advanced surveillance technologies that have been developed—and are likely to be developed—including:

- *High-power zoom lenses.* UAVs can carry increasingly powerful lenses that allow significant zooming, increasing the chance that individuals will come under scrutiny from faraway aircraft without knowing it. And the density of photo sensors is growing at an exponential pace (in line with Moore's law), allowing for higher and higher resolution photos to be taken for the same price camera.[14]

- *Night vision.* Infrared and ultraviolet imaging enable night vision by capturing light outside the spectrum visible to the human eye. Infrared imaging (also known as thermal imaging) shows heat emitted by an object, and so is especially suited for identifying humans and animals in the dark.[15] Ultraviolet (UV) imaging can detect some materials not visible in natural or infrared light, and can also be used to enhance detail; for instance, it can be used to image surface textures not apparent in visible light.[16] Moving forward, thermal imaging is likely to improve—for example becoming more sensitive and available at higher resolutions.
- *See-through imaging.* The military is developing radar technologies that can see through ceilings and walls and allow the tracking of human targets even when they are inside buildings.[17] A technology called Synthetic Aperture Radar, for example, can see through cloudy and dusty conditions and through foliage, and has the potential to penetrate the earth and walls.[18]
- Video analytics. This field seeks to apply artificial intelligence techniques not just to collect but also to "watch" video. The technology has been improving rapidly, and can recognize and respond to specific people, events, and objects.[19] One of the most significant uses would be to continually track individuals or vehicles as they move about, using face recognition or other bodily characteristics.[20] It might also be used to identify particular movement patterns as "suspicious," or to identify and flag changes in routines, buildings or grounds.[21] Computers performing these tasks have a distinct advantage over human observers, because as one observer summed it up, "machines do not blink or forget. They are tireless assistants."[22]
- The PBS series NOVA, "Rise of the Drones," recently aired a segment detailing the capabilities of a powerful aerial surveillance system known as ARGUS-IS. This system, which is basically a super-high, 1.8 gigapixel resolution camera that can be mounted on a drone, demonstrates many of these capacities. The system is capable of high-resolution monitoring and recording of an entire city. To witness a

demonstration of this capacity please see: http://www.youtube.com/watch?feature=playerembedded&v=13Bahr dkMU8

IV. UAVs and Possible Harms

With the federal government likely to permit more widespread use of drones, and the technology likely to become ever more powerful, the question becomes: what role will drones play in American life? Based on current trends—technology development, law enforcement interest, political and industry pressure, and the lack of legal safeguards—it is clear that drones pose a looming threat to Americans' privacy. The reasons for concern reach across a number of different dimensions:

- *Mission creep.* Even where UAVs are being envisioned for search and rescue, fighting wildfires, and in dangerous tactical police operations, they are likely to be quickly embraced by law enforcement around the nation for other, more controversial purposes. The police in Ogden, Utah think that floating a surveillance blimp above their city "will be a deterrent to crime when it is out and about."[23] In Houston, police suggested that drones could possibly be used for writing traffic tickets. [24] The potential result is that they become commonplace in American life.
- *Tracking.* The Justice Department currently claims the authority to monitor Americans' comings and goings using cell phone and GPS tracking devices—under uncertain legal standards. Fleets of UAVs, interconnected and augmented with analytics software, could enable the mass tracking of vehicles and pedestrians around a wide area.
- *New uses.* The use of drones could also be expanded from surveillance to actual intervention in law enforcement situations on the ground. Airborne technologies could be developed that could, for example, be used to control or dispel protesters (perhaps by deploying tear gas or other technologies), stop a fleeing vehicle, or even deploy weapons.[25]

In addition, drones raise many of the same issues that pervasive video surveillance brings in any context. For example:

- *Chilling effects.* What would be the effect on our public spaces, and our society as a whole, if everyone felt the keen eye of the government on their backs whenever they ventured outdoors? Psychologists have repeatedly found that people who are being observed tend to behave differently, and make different decisions, than when they are not being watched. This effect is so great that a recent study found that "merely hanging up posters of staring human eyes is enough to significantly change people's behavior."[26]
- *Voyeurism.* Video surveillance is susceptible to individual abuse, including voyeurism. In 2004, a couple making love on a dark nighttime rooftop balcony, where they had every reason to expect they enjoyed privacy, were filmed for nearly four minutes by a New York police helicopter using night vision. This is the kind of abuse that could become commonplace if drone technology enters widespread use. (Rather than apologize, NYPD officials flatly denied that this filming constituted an abuse, telling a television reporter, "this is what police in helicopters are supposed to do, check out people to make sure no one is ... doing anything illegal").[27]
- *Discriminatory targeting.* The individuals operating surveillance systems bring to the job all their existing prejudices and biases. In Great Britain, camera operators have been found to focus disproportionately on people of color. According to a sociological study of how the systems were operated, "Black people were between one-and-a-half and two-and-a-half times more likely to be surveilled than one would expect from their presence in the population."[28]
- *Institutional abuse.* In addition to abuse by the inevitable "bad apples" within law enforcement, there is also the danger of institutional abuse. Sometimes, bad policies are set at the top, and an entire law enforcement agency is turned toward abusive ends. That is especially prone to happen in periods of social turmoil and intense political conflict. During the labor, civil rights, and anti-Vietnam war movements of the 20th century, the FBI and other security agencies engaged in systematic illegal behavior against those challenging the status quo. And once again today we are seeing an upsurge in spying against peaceful political protesters across America.[29]
- *Automated enforcement.* Drones are part of a trend toward automated law enforcement, in which cameras and other technologies are used to mete out justice with little or no human intervention. This trend raises a variety of concerns, such as the fact

> that computers lack the judgment to fairly evaluate the circumstances surrounding a supposed violation, and may be susceptible to bugs and other software errors, or simply are not programmed to fairly and properly encapsulate the state of the law as passed by legislatures.[30]

One point that is often made about new surveillance technologies is that, while they may increase government surveillance of individuals, they can also increase individuals' ability to record the activities of officials, which can serve as a check on their power. Too often, however, the authorities seek to increase their surveillance over individuals (for example, by installing surveillance cameras throughout public spaces) while restricting individuals' ability to use that same technology as a check against their power (for example, by attempting to prevent individuals from videotaping police[31]). Already, security experts have started expressing concern that unmanned aircraft could be used for terrorism[32]—which naturally raises the question: will individuals be able to make use of the new technology for their own purposes, or will government seek a monopoly over the new technology by citing fears of its use for terrorism?

V. The Fourth Amendment and the Use of Drones

The Supreme Court has never taken a position on whether the Fourth Amendment places limits on government use of UAV surveillance. However, it allowed some warrantless aerial surveillance from *manned* aircraft.

- In the 1986 decision *California v. Ciraolo*, the Supreme Court focused on whether an individual has a privacy interest in being free from aerial surveillance of his backyard. The police had received a tip that Dante Ciraolo was growing marijuana in his backyard, but high fences prevented them from viewing his backyard from the street. The police borrowed a plane, flew it over the backyard and easily spotted marijuana plants growing there. Ciraolo argued that his Fourth Amendment rights were violated because the government did not get a warrant. The Court rejected this argument, explaining that there was no intrusion into his privacy because "[a]ny member of the public flying in this airspace who glanced down could have seen everything that these officers observed."[33]

- *Dow Chemical Co. v. United States*, also decided in 1986, the Supreme Court addressed whether the Environmental Protection Agency violated Dow's Fourth Amendment rights when it employed a commercial aerial photographer to use a precision aerial mapping camera to take photographs of a chemical plant. The Court found no violation, in part because the camera the EPA used was a "conventional, albeit precise, commercial camera commonly used in mapmaking," and "the photographs here are not so revealing of intimate details as to raise constitutional concerns." However, the Court suggested that the use of more sophisticated, intrusive surveillance might justify a different result. It wrote, "surveillance of private property by using highly sophisticated surveillance equipment not generally available to the public, such as satellite technology, might be constitutionally proscribed absent a warrant."[34]
- In *Florida v. Riley*, decided in 1989, the police had received a tip that Michael Riley was growing marijuana in a greenhouse on the property surrounding his home. The interior of the greenhouse was not visible from the ground outside the property, and the greenhouse had a ceiling, though two panels in the ceiling were missing. A police officer flew over the greenhouse and spotted marijuana through the openings in the roof. While no reasoning commanded a majority of the Court, four justices concluded that its decision in *Ciraolo* applied because Riley had left part of the greenhouse open to public view, and so the search was constitutional.[35]

Because of their potential for pervasive use in ordinary law enforcement operations and capacity for revealing far more than the naked eye, drones pose a more serious threat to privacy than do manned flights. There are good reasons to believe that they may implicate Fourth Amendment rights in ways that manned flights do not.

Government use of UAVs equipped with technology that dramatically improves on human vision or captures something humans cannot see (such thermal or x-ray images) should be scrutinized especially closely by the courts. This follows from the Supreme Court's statement in Dow Chemical that using sophisticated technology not generally available to the public may be considered a search under the Fourth Amendment. It is also suggested by the

2001 case *Kyllo v. United States*, in which the court rejected the use of thermal imaging devices to peer into a suspect's home without a warrant.[36]

Further, the Supreme Court has suggested that the pervasive or continuous use of a surveillance technology may heighten Fourth Amendment concerns. In *United States v. Knotts*, the Supreme Court addressed whether attaching primitive "beeper" tracking technology to a car violated the driver's Fourth Amendment rights.[37] Although it concluded that the use of the beeper in that case did not violate the Fourth Amendment, it held that if "such dragnet type law enforcement practices" as "twenty-four hour surveillance of any citizen of this country" ever arose, it would determine if different constitutional principles would be applicable.

Similarly, in *US v. Jones*, decided last year, a concurrence joined by 5 justices found that GPS tracking of a car implicated an individual's reasonable expectation of privacy and noted "society's expectation has been that law enforcement agents and others would not—and indeed, in the main, simply could not—secretly monitor and catalogue every single movement of an individual's car for a very long period."[38] While this decision may eventually play a role in regulating drone usage, the technology is moving far more rapidly than our jurisprudence, and it is critical that Congress not delay action, especially with a looming 2015 deadline set by the FAA Reauthorization Act.

VI. Recommendations

UAVs can be an extremely powerful surveillance tool, and their use must be subject to strict limitations, as should all government power. Like any tool, UAVs have the potential to be used for good or ill. With implementation of good privacy ground rules, our society can enjoy the benefits of this technology without having to worry about its darker potential. Placing reasonable limitations on law enforcement is by no means a new idea. For example authorities may take a thermal image of someone's home only when they get a warrant. Congress should impose appropriate rules, limits and regulations on UAVs as well in order to preserve the privacy Americans have always expected and enjoyed.

At a minimum, Congress should enact the following core measures to ensure that this happens:

- Usage restrictions. UAVs should be subject to strict regulation to ensure that their use does not eviscerate the privacy that Americans have traditionally enjoyed and rightly expect. Innocent Americans should not have to worry that their activities will be scrutinized by drones. To this end, the use of drones should be prohibited for indiscriminate mass surveillance, for example, or for spying based on First Amendment-protected activities. In general, drones should not be deployed except:

- where there are specific and articulable grounds to believe that the drone will collect evidence relating to a specific instance of criminal wrongdoing or, if the drone will intrude upon non-public spaces where the government has obtained a warrant based on probable cause; or
- where there is a geographically confined, time-limited emergency situation in which particular individuals' lives are at risk, such as a fire, hostage crisis, or person lost in the wilderness; or
- for reasonable non-law enforcement purposes by non-law enforcement agencies, where privacy will not be substantially affected, such as geological inspections or environmental surveys, and where the surveillance will not be used for secondary law enforcement purposes.

- *Image retention restrictions.* Images of identifiable individuals captured by aerial surveillance technologies should not be retained or shared unless there is reasonable suspicion that the images contain evidence of criminal activity or are relevant to an ongoing investigation or pending criminal trial.
- *Public notice.* The policies and procedures for the use of aerial surveillance technologies should be explicit and written, and should be subject to public review and comment. While it is legitimate for the police to keep the details of particular investigations confidential, policy decisions regarding overall deployment policies—including the privacy trade-offs they may entail—are a public matter that should be openly discussed.
- *Democratic control.* Deployment and policy decisions surrounding UAVs should be democratically decided based on open information—not made on the fly by police departments simply by virtue of federal grants or other autonomous purchasing decisions or departmental policy fiats.
- *Auditing and effectiveness tracking.* Investments in UAVs should only be made with a clear, systematic examination of the costs and benefits involved. And if aerial surveillance technology is deployed,

independent audits should be put in place to track the use of UAVs by government, so that citizens and other watchdogs can tell generally how and how often they are being used, whether the original rationale for their deployment is met, whether they represent a worthwhile public expenditure, and whether they are being used for improper or expanded purposes.
- *Ban on weaponization.* Weapons developed on the battlefield in Iraq and Afghanistan have no place inside the U.S. The national consensus on this issue is reflected by the fact that the Heritage Foundation and the International Association of Chiefs of Police join us in supporting sharp limits on weaponized drones.[39]

State legislatures are already responding to the need to safeguard against drone surveillance. According to the National Conference of State Legislatures, during the current legislative session 42 states have considered 95 different bills and resolutions concerning drones.[40] The vast majority of these bills are focused squarely on privacy issues associated with drone use. Already, Florida[41], Idaho[42], Montana[43], and Virginia[44] have enacted drones legislation; a bill awaits the governor's approval in Tennessee[45]; and legislation has passed one chamber in Illinois[46], Missouri[47], North Dakota[48], and Texas.[49] The amount of progress on drones this legislative session is impressive given that this is the first session in which state legislatures have considered the issue. The astonishing level of activity strongly indicates just how concerned state legislators and their constituents are about ensuring that any drone use respects individuals' privacy rights and expectations.

The House Judiciary Committee has before it a very strong and privacy-protective solution to the problems created by drone use: H.R. 637, the Preserving American Privacy Act of 2013. This bipartisan measure, sponsored by Rep. Ted Poe (R-TX) and co-sponsored by numerous other members of the Judiciary Committee, sharply limits the potential misuse of drones by the government while still allowing responsible private sector use of the technology. It requires judicial approval for all drone flights by the government, limits unnecessary information collection, creates a suppression remedy for wrongly collected evidence, bans the weaponization of drones and establishes reporting requirements for drone use. Under the legislation the Attorney General may also withdraw licensing for government drones if they operate outside of allowable parameters.

Drone technology certainly has beneficial uses – for search and rescue missions, firefighting, dangerous police tactical operations – but also poses significant possible harms if left unchecked. Ultimately this powerful new technology should only be used if subject to an equally powerful framework that regulates its use in order to avoid abuse and invasions of privacy.

End Notes

[1] FAA Modernization and Reform Act of 2012, P.L. 112-95, §332, 126 Stat.11, 73. For privacy controls the ACLU recommends the FAA implement as part of its regulatory process please see: http://www.aclu.org/technology-andliberty/aclu-comment-faa-unmanned-aircraft-system-test-site-program.

[2] See Wikipedia, "List of unmanned aerial vehicles," at http://en.wikipedia.org/wiki/List_of_unmanned_aerial_vehicles.

[3] "Israel unveils world's largest UAV," Homeland Security Newswire, Feb. 23, 2010, online at http://homelandsecuritynewswire.com/israel-unveils-worlds-largest-uav.

[4] See General Atomics web page on Predator B at http://www.ga-asi.com/products/aircraft/predator_b.php; R.P.G. Collinson, Introduction to Avionic Systems (2011), p. 495

[5] Yochi J. Dreazen, "From Pakistan, With Love: The technology used to monitor the skies over Waziristan is coming to your hometown," National Journal, March 13, 2011, online at http://www.nationaljournal.com/magazine/drones-may-be-coming-to-your-hometown-20110313.

[6] Stephen Dean, "Police line up to use drones on patrol after Houston secret test," Houston Examiner, Jan. 11, 2010, online at http://www.examiner.com/page-one-in-houston/police-line-up-to-use-drones-on-patrol-after-houstonsecret-test.

[7] Insitu, ScanEagle brochure, online at http://www.insitu.com/documents/Insitu%20Website/Marketing%20Collateral/ScanEagle%20Folder%20Insert.pdf

[8] AeroVironment brochure, online at http://www.avinc.com/downloads/Raven_ Domestic_ 1210.pdf; AeroVironment web page on the Wasp at http://www.avinc.com/uas/small_uas/wasp/; Carrie Kahn, "It's A Bird! It's A Plane! It's A Drone!" National Public Radio, March 14, 2011, online at http://www.npr.org/2011/03/14/134533552/its-a-bird-itsa-plane-its-a-drone; "Drones on the home front," Washington Post, Jan. 23, 2011, online at http://www.washingtonpost.com/wp-srv/special/nation/drone-gallery/.

[9] W.J. Hennigan, "It's a bird! It's a spy! It's both," Los Angeles Times, Feb. 17, 2011, online at http://articles.latimes.com/2011/feb/17/business/la-fi-hummingbird-drone-20110217.

[10] On high-altitude blimps see Elliott Minor, "Interest Growing in 'Security' Blimps," Associated Press, April 27, 2004, available online at http://www.rustysforum.com/cgi-bin/domains/com/rustysforum/frc_bb/ultimatebb.cgi?ubb=next_topic&f=1&t=000807&go=older; on low-altitude blimps see e.g. James Nelson, "Utah city may use blimp as anti-crime spy in the sky," Reuters, Jan. 16, 2011, online at http://www.reuters.com/article/2011/01/16/us-crime-blimp-utah-idUSTRE70F1DJ20110116.

[11] W.J. Hennigan, "It's a bird! It's a spy! It's both," Los Angeles Times, Feb. 17, 2011, online at http://articles.latimes.com/2011/feb/17/business/la-fi-hummingbird-drone-20110217.

[12] M. Ryan Calo, "Robots and Privacy," April 2010, online at http://ssrn.com/abstract=1599189.

[13] "Gliders Emerge As Surveillance UAVs," Aviation Week, June 8, 2010, online at http://www.aviationweek.com/aw/generic/story_generic.jsp?topicName=ila_2010&id=news/awx/2010/06/08/awx_0 6_08_2010j0-232627.xml; James Nelson, "Utah city may use

blimp as anti-crime spy in the sky," Reuters, Jan. 16, 2011, online at http://www.reuters.com/article/2011/01/16/us-crime-blimp-utah-idUSTRE70F1DJ20110116; Ned Smith, "Solar-powered UAV can stay aloft 5 years," TechNewsDaily, Sept. 22, 2010, online at http://www.msnbc.msn.com/id/39313306/ns/technology_and_science-tech_and_gadgets/t/solar-powered-uav-canstay-aloft-years.

[14] Nathan Myhrvold, "Moore's Law Corollary: Pixel Power," New York Times, June 7, 2006, online at http://www.nytimes.com/2006/06/07/technology/circuits/07essay.html. Moore's law is the observation that the number of transistors that can be placed on an integrated circuit—and therefore broadly speaking the power of computers—doubles approximately every two years. It has held true for over 50 years.

[15] NASA Science Mission Directorate, "Infrared Energy," Mission: Science, 2010, online at http://missionscience.nasa.gov/ems/07_infraredwaves.html.

[16] Austin Richards, "Digital Reflected-Ultraviolet Imaging," Advanced Imaging, Apr. 2006, online at http://www.uvcorder.com/pdf/ADI0406%20Component%2018-20.pdf.

[17] See e.g., William Saletan, "Nowhere To Hide," Slate.com, Sept. 17, 2008, online at http://www.slate.com/articles/health_and_science/human_nature/2008/09/nowhere_to_hide.html Greg Miller and Julian E. Barnes, "Special drones pursue militias," Los Angeles Times, Sept. 12, 2008, online at http://articles.latimes.com/2008/sep/12/world/fg-pakistan 12.

[18] "Ground Moving Target Indicator (GMTI) Radar Discrimination of Combatants versus Animals in Severe Clutter," DARPA, undated document (topic number SB082-019), online at http://www.dodsbir.net/sitis/archives_display_topic.asp?Bookmark=32303. Sandia National Laboratories, "Synthetic Aperture Radar Applications," undated, online at http://www.sandia.gov/radar/sarapps.html; Alicia Tejada, "MIT Develops New Radar Technology: Military Could See Through Walls," ABC News, Oct. 20, 2011, online at http://abcnews.go.com/Technology/radar-technology-mit-walls/story?id=14773871.

[19] Vigilant Video, online at http://www.vigilantvideo.com.

[20] Noah Shachtman, "Army Tracking Plan: Drones That Never Forget a Face," Wired.com, Sept. 28, 2011, online at http://www.wired.com/dangerroom/2011/09/drones-never-forget-a-face/.

[21] On change detection, see Sandia National Laboratories, "Synthetic Aperture Radar Applications," undated, online at http://www.sandia.gov/radar/sarapps.html.

[22] Steve Lohr, "Computers That See You and Keep Watch Over You," New York Times, Jan. 1, 2011, online at http://www.nytimes.com/2011/01/02/science/02see.html.

[23] James Nelson, "Utah city may use blimp as anti-crime spy in the sky," Reuters, Jan. 16, 2011, online at http://www.reuters.com/article/2011/01/16/us-crime-blimp-utah-idUSTRE70 F1DJ20110116.

[24] Stephen Dean, "Police line up to use drones on patrol after Houston secret test," Houston Examiner, Jan. 11, 2010, online at http://www.examiner.com/page-one-in-houston/police-line-up-to-use-drones-on-patrol-after-houstonsecret-test.

[25] Joseph Nevins, "Robocop: Drones at Home," Boston Review, January/February 2011, online at http://www.bostonreview.net/BR36.1/nevins.php.

[26] Sander van der Linden, "How the Illusion of Being Observed Can Make You a Better Person," Scientific American, May 3, 2011, online at http://www.scientificamerican.com/article.cfm?id=how-the-illusion-of-beingobserved-can-make-you-better-person; M. Ryan Calo, "People Can Be So Fake: A New Dimension to Privacy and Technology Scholarship," 114 Penn St. L. Rev. 809, online at http://www.pennstatelawreview.org/articles/114/114%20Penn%20St.%20L.%20Rev.%20809.pdf.

[27] "Did NYPD Cameras Invade A Couple's Privacy?" WCBS-TV report, Feb. 24, 2005, video no longer available online; Jim Dwyer, "Police Video Caught a Couple's Intimate Moment on a Manhattan Rooftop," New York Times, Dec. 22, 2005, online at http://www.nytimes.com/2005/12/22/nyregion/22rooftop.html.

[28] Clive Norris and Gary Armstrong, "The Unforgiving Eye: CCTV Surveillance in Public Spaces," Centre for Criminology and Criminal Justice at Hull University, 1997.

[29] See ACLU "Spyfiles" web site at www.aclu.org/spyfiles.

[30] Danielle Keats Citron, "Technological Due Process," 85 Washington University Law Review 1249 (2008), online at http://lawreview.wustl.edu/inprint/85/6/Citron.pdf.

[31] See Jay Stanley, "You Have Every Right to Photograph That Cop," ACLU, online at http://www.aclu.org/freespeech/you-have-every-right-photograph-cop.

[32] Agence France Press, "Flying Robot Attacks 'Unstoppable' Say Experts," Agence France Press, May 11, 2006, available online at http://www.rense.com/general71/sspm.htm.

[33] 476 U.S. 207 (1986).

[34] 476 U.S. 227 (1986).

[35] 488 U.S. 445 (1989).

[36] 533 U.S. 27 (2001).

[37] 460 U.S. 276, 283-84 (1983).

[38] 132 S.Ct. 945.

[39] International Assocation of Cheifs of Police, Aviation Committee, Recommended Guidelines for the use of Unmanned Aircraft. August 2012, see: http://www.theiacp.org/portals /0/pdfs/IACP_UAGuidelines.pdf; Paul Rosenzweig, Steven P. Bucci, Ph.D., Charles "Cully" Stimson and James Jay Carafano, Ph.D., Drones in U.S. Airspace: Principles for Governance, The Heritage Foundation, September 20, 2012, see: http://www.heritage.org/research/reports/2012/09/drones-in-us-airspace-principles-for-governance.

[40] National Conference of State Legislatures, 2013 Unmanned Aircraft Systems (UAS) Legislation, http://www.ncsl.org/issues-research/justice/unmanned-aerial-vehicles.aspx.

[41] http://flsenate.gov/Session/Bill/2013/0092/BillText/er/HTML.

[42] http://www.legislature.idaho.gov/legislation/2013/S1134.htm.

[43] http://data.opi.mt.gov/bills/2013/billpdf/SB0196.pdf.

[44] http://leg1.state.va.us/cgi-bin/legp504.exe?131+ful+CHAP0755.

[45] http://wapp.capitol.tn.gov/apps/BillInfo/Default.aspx?BillNumber=SB0796.

[46] http://www.ilga.gov/legislation/BillStatus.asp?DocNum=1587&GAID=12&DocTypeID.= SB &LegID=72407&Sessi onID=85&SpecSess=&Session=&GA=98.

[47] http://www.house.mo.gov/billsummary.aspx?bill=HB46&year=2013&code=R.

[48] http://www.legis.nd.gov/assembly/63-2013/bill-actions/ba1373.html.

[49] http://www.capitol.state.tx.us/BillLookup/Actions.aspx?LegSess=83R&Bill=HB912.

In: Drones in the U.S.
Editor: Sophia Barrett

ISBN: 978-1-63484-918-0

Chapter 8

PRESIDENTIAL MEMORANDUM: PROMOTING ECONOMIC COMPETITIVENESS WHILE SAFEGUARDING PRIVACY, CIVIL RIGHTS, AND CIVIL LIBERTIES IN DOMESTIC USE OF UNMANNED AIRCRAFT SYSTEMS*

Barack Obama

MEMORANDUM FOR THE HEADS OF EXECUTIVE DEPARTMENTS AND AGENCIES

SUBJECT: Promoting Economic Competitiveness While Safeguarding Privacy, Civil Rights, and Civil Liberties in Domestic Use of Unmanned Aircraft Systems

Unmanned Aircraft Systems (UAS) technology continues to improve rapidly, and increasingly UAS are able to perform a variety of missions with greater operational flexibility and at a lower cost than comparable manned aircraft. A wide spectrum of domestic users -- including industry, private citizens, and Federal, State, local, tribal, and territorial governments -- are

* This is an edited, reformatted and augmented version of a presidential memorandum issued by the White House, Office of the Press Secretary, February 15, 2015.

using or expect to use these systems, which may play a transformative role in fields as diverse as urban infrastructure management, farming, public safety, coastal security, military training, search and rescue, and disaster response.

The Congress recognized the potential wide-ranging benefits of UAS operations within the United States in the FAA Modernization and Reform Act of 2012 (Public Law 112-95), which requires a plan to safely integrate civil UAS into the National Airspace System (NAS) by September 30, 2015. As compared to manned aircraft, UAS may provide lower-cost operation and augment existing capabilities while reducing risks to human life. Estimates suggest the positive economic impact to U.S. industry of the integration of UAS into the NAS could be substantial and likely will grow for the foreseeable future.

As UAS are integrated into the NAS, the Federal Government will take steps to ensure that the integration takes into account not only our economic competitiveness and public safety, but also the privacy, civil rights, and civil liberties concerns these systems may raise.

By the authority vested in me as President by the Constitution and the laws of the United States of America, and in order to establish transparent principles that govern the Federal Government's use of UAS in the NAS, and to promote the responsible use of this technology in the private and commercial sectors, it is hereby ordered as follows:

Section 1. UAS Policies and Procedures for Federal Government Use. The Federal Government currently operates UAS in the United States for several purposes, including to manage Federal lands, monitor wildfires, conduct scientific research, monitor our borders, support law enforcement, and effectively train our military. As with information collected by the Federal Government using any technology, where UAS is the platform for collection, information must be collected, used, retained, and disseminated consistent with the Constitution, Federal law, and other applicable regulations and policies. Agencies must, for example, comply with the Privacy Act of 1974 (5 U.S.C. 552a) (the "Privacy Act"), which, among other things, restricts the collection and dissemination of individuals' information that is maintained in systems of records, including personally identifiable information (PII), and permits individuals to seek access to and amendment of records.

(a) Privacy Protections. Particularly in light of the diverse potential uses of UAS in the NAS, expected advancements in UAS technologies, and the anticipated increase in UAS use in the future, the Federal Government shall take steps to ensure that privacy protections and policies relative to UAS continue to keep pace with these developments. Accordingly, agencies shall,

prior to deployment of new UAS technology and at least every 3 years, examine their existing UAS policies and procedures relating to the collection, use, retention, and dissemination of information obtained by UAS, to ensure that privacy, civil rights, and civil liberties are protected. Agencies shall update their policies and procedures, or issue new policies and procedures, as necessary. In addition to requiring compliance with the Privacy Act in applicable circumstances, agencies that collect information through UAS in the NAS shall ensure that their policies and procedures with respect to such information incorporate the following requirements:

(i) Collection and Use. Agencies shall only collect information using UAS, or use UAS-collected information, to the extent that such collection or use is consistent with and relevant to an authorized purpose.

(ii) Retention. Information collected using UAS that may contain PII shall not be retained for more than 180 days unless retention of the information is determined to be necessary to an authorized mission of the retaining agency, is maintained in a system of records covered by the Privacy Act, or is required to be retained for a longer period by any other applicable law or regulation.

(iii) Dissemination. UAS-collected information that is not maintained in a system of records covered by the Privacy Act shall not be disseminated outside of the agency unless dissemination is required by law, or fulfills an authorized purpose and complies with agency requirements.

(b) Civil Rights and Civil Liberties Protections. To protect civil rights and civil liberties, agencies shall:

(i) ensure that policies are in place to prohibit the collection, use, retention, or dissemination of data in any manner that would violate the First Amendment or in any manner that would discriminate against persons based upon their ethnicity, race, gender, national origin, religion, sexual orientation, or gender identity, in violation of law;

(ii) ensure that UAS activities are performed in a manner consistent with the Constitution and applicable laws, Executive Orders, and other Presidential directives; and

(iii) ensure that adequate procedures are in place to receive, investigate, and address, as appropriate, privacy, civil rights, and civil liberties complaints.

(c) Accountability. To provide for effective oversight, agencies shall:

(i) ensure that oversight procedures for agencies' UAS use, including audits or assessments, comply with existing agency policies and regulations;

(ii) verify the existence of rules of conduct and training for Federal Government personnel and contractors who work on UAS programs, and procedures for reporting suspected cases of misuse or abuse of UAS technologies;

(iii) establish policies and procedures, or confirm that policies and procedures are in place, that provide meaningful oversight of individuals who have access to sensitive information (including any PII) collected using UAS;

(iv) ensure that any data-sharing agreements or policies, data use policies, and record management policies applicable to UAS conform to applicable laws, regulations, and policies;

(v) establish policies and procedures, or confirm that policies and procedures are in place, to authorize the use of UAS in response to a request for UAS assistance in support of Federal, State, local, tribal, or territorial government operations; and

(vi) require that State, local, tribal, and territorial government recipients of Federal grant funding for the purchase or use of UAS for their own operations have in place policies and procedures to safeguard individuals' privacy, civil rights, and civil liberties prior to expending such funds.

(d) Transparency. To promote transparency about their UAS activities within the NAS, agencies that use UAS shall, while not revealing information that could reasonably be expected to compromise law enforcement or national security:

(i) provide notice to the public regarding where the agency's UAS are authorized to operate in the NAS;

(ii) keep the public informed about the agency's UAS program as well as changes that would significantly affect privacy, civil rights, or civil liberties; and

(iii) make available to the public, on an annual basis, a general summary of the agency's UAS operations during the previous fiscal year, to include a brief description of types or categories of missions flown, and the number of times the agency provided assistance to other agencies, or to State, local, tribal, or territorial governments.

(e) Reports. Within 180 days of the date of this memorandum, agencies shall provide the President with a status report on the implementation of this section. Within 1 year of the date of this memorandum, agencies shall publish information on how to access their publicly available policies and procedures implementing this section.

Sec. 2. Multi-stakeholder Engagement Process. In addition to the Federal uses of UAS described in section 1 of this memorandum, the combination of greater operational flexibility, lower capital requirements, and lower operating costs could allow UAS to be a transformative technology in the commercial and private sectors for fields as diverse as urban infrastructure management, farming, and disaster response. Although these opportunities will enhance American economic competitiveness, our Nation must be mindful of the potential implications for privacy, civil rights, and civil liberties. The Federal Government is committed to promoting the responsible use of this technology in a way that does not diminish rights and freedoms.

(a) There is hereby established a multi-stakeholder engagement process to develop and communicate best practices for privacy, accountability, and transparency issues regarding commercial and private UAS use in the NAS. The process will include stakeholders from the private sector.

(b) Within 90 days of the date of this memorandum, the Department of Commerce, through the National Telecommunications and Information Administration, and in consultation with other interested agencies, will initiate this multi-stakeholder engagement process to develop a framework regarding privacy, accountability, and transparency for commercial and private UAS use. For this process, commercial and private use includes the use of UAS for commercial purposes as civil aircraft, even if the use would qualify a UAS as a public aircraft under 49 U.S.C. 40102(a)(41) and 40125. The process shall not focus on law enforcement or other noncommercial governmental use.

Sec. 3. Definitions. As used in this memorandum:

(a) "Agencies" means executive departments and agencies of the Federal Government that conduct UAS operations in the NAS.

(b) "Federal Government use" means operations in which agencies operate UAS in the NAS. Federal Government use includes agency UAS operations on behalf of another agency or on behalf of a State, local, tribal, or territorial government, or when a nongovernmental entity operates UAS on behalf of an agency.

(c) "National Airspace System" means the common network of U.S. airspace; air navigation facilities, equipment, and services; airports or landing areas; aeronautical charts, information, and services; related rules, regulations, and procedures; technical information; and manpower and material. Included in this definition are system components shared jointly by the Departments of Defense, Transportation, and Homeland Security.

(d) "Unmanned Aircraft System" means an unmanned aircraft (an aircraft that is operated without direct human intervention from within or on the aircraft) and associated elements (including communication links and components that control the unmanned aircraft) that are required for the pilot or system operator in command to operate safely and efficiently in the NAS.

(e) "Personally identifiable information" refers to information that can be used to distinguish or trace an individual's identity, either alone or when combined with other personal or identifying information that is linked or linkable to a specific individual, as set forth in Office of Management and Budget Memorandum M-07-16 (May 22, 2007) and Office of Management and Budget Memorandum M-10-23 (June 25, 2010).

Sec. 4. General Provisions.

(a) This memorandum complements and is not intended to supersede existing laws and policies for UAS operations in the NAS, including the National Strategy for Aviation Security and its supporting plans, the FAA Modernization and Reform Act of 2012, the Federal Aviation Administration's (FAA's) Integration of Civil UAS in the NAS Roadmap, and the FAA's UAS Comprehensive Plan.

(b) This memorandum shall be implemented consistent with applicable law, and subject to the availability of appropriations.

(c) Nothing in this memorandum shall be construed to impair or otherwise affect:

(i) the authority granted by law to an executive department, agency, or the head thereof; or

(ii) the functions of the Director of the Office of Management and Budget relating to budgetary, administrative, or legislative proposals.

(d) Independent agencies are strongly encouraged to comply with this memorandum.

(e) This memorandum is not intended to, and does not, create any right or benefit, substantive or procedural, enforceable at law or in equity by any party against the United States, its departments, agencies, or entities, its officers, employees, or agents, or any other person.

(f) The Secretary of Commerce is hereby authorized and directed to publish this memorandum in the Federal Register.

BARACK OBAMA

INDEX

#

A

B

C

D

E

F

G

H

I

J

K

L

M

N

O

P

R

S